# From the Corn Belt to the Gulf

## Societal and Environmental Implications of Alternative Agricultural Futures

*Edited by*
**Joan Iverson Nassauer,**
**Mary V. Santelmann, and Donald Scavia**

Resources for the Future
Washington, DC, USA

Printed in the United States of America

An RFF Press book
Published by Resources for the Future
1616 P Street NW
Washington, DC 20036–1400
USA
www.rffpress.org

*Library of Congress Cataloging-in-Publication Data*

From the corn belt to the gulf : societal and environmental implications of alternative agricultural futures / edited by Joan Iverson Nassauer, Mary V. Santelmann, and Donald Scavia. -- 1st ed.
     p. cm.
  ISBN 978-1-933115-47-4 (hardcover : alk. paper) -- ISBN 978-1-933115-48-1 (pbk. : alk. paper) 1. Agriculture--Environmental aspects--Mississippi River Watershed. 2. Agriculture--Environmental aspects--Mexico, Gulf of. 3. Agriculture and state--Mississippi River Watershed. 4. Agriculture and state--Mexico, Gulf of. I. Nassauer, Joan Iverson. II. Santelmann, Mary V. III. Scavia, Donald. IV. Title: Societal and environmental implications of alternative agricultural futures.
  S589.757.M57F76 2007
  338.1′620977--dc22
                            2007017250

The paper in this book meets the guidelines for permanence and durability of the Committee on Production Guidelines for Book Longevity of the Council on Library Resources. This book was typeset by Jeff Hall. It was copyedited by René Howard. The cover was designed by Maggie Powell. Cover photos by Chris Faust.

ISBN  978-1-933115-47-4  (cloth)          ISBN  978-1-933115-48-1 (paper)

# *About* Resources for the Future *and* RFF Press

**Resources for the Future (RFF)** improves environmental and natural resource policymaking worldwide through independent social science research of the highest caliber. Founded in 1952, RFF pioneered the application of economics as a tool for developing more effective policy about the use and conservation of natural resources. Its scholars continue to employ social science methods to analyze critical issues concerning pollution control, energy policy, land and water use, hazardous waste, climate change, biodiversity, and the environmental challenges of developing countries.

**RFF Press** supports the mission of RFF by publishing book-length works that present a broad range of approaches to the study of natural resources and the environment. Its authors and editors include RFF staff, researchers from the larger academic and policy communities, and journalists. Audiences for publications by RFF Press include all of the participants in the policymaking process—scholars, the media, advocacy groups, NGOs, professionals in business and government, and the public.

# Contents

## Part 1: Environmental and Societal Drivers of Agricultural Landscape Futures

## Part 2: Assessing the Effects of Alternative Corn Belt Landscape Futures in Iowa

## Part 3: Policy Implications across Scales: From Iowa Watersheds to the Mississippi River Basin

# Dedication

We dedicate this book to Roger Iverson and Lyle Santelmann, and all the farmers of the Corn Belt who taught us to see its changing landscape.

And to the memory of George W. Carver, the Iowa State University agriculturalist whose insights and vision transformed the agricultural landscapes of his time.

# Acknowledgments

This book grew from the interdisciplinary collaboration of two groups: the Corn Belt futures research team, which developed and assessed alternative scenarios and future landscapes for two small agricultural watersheds in Iowa, and the scientists supporting the intergovernmental Task Force on Hypoxia in the Gulf of Mexico, which examined how land use across the entire Mississippi River Basin (MRB) might affect the "dead zone" in the Gulf of Mexico. Each of these groups depended on the efforts of many people and drew on past successes of many others to inform their efforts. Our attempts to acknowledge these contributions here will inevitably be incomplete, and we can only begin this book with gratitude that so many others have also found it worth their time to observe, engage, experiment, and invest in the future of agricultural landscapes and their link to environmental health.

The Corn Belt futures team formally began work in 1996, when Kathryn Freemark Lindsay, Joan Nassauer, Mary Santelmann, and Denis White led development of a proposal in response to the Environmental Protection Agency and National Science Foundation Water and Watersheds program.[1] As we developed the proposal, we began to envision a truly interdisciplinary project in which we combined our separate types of science to explore how agricultural landscapes could change for the better – for their effects on the environment and on quality of life. Several of us had previously studied the intensively farmed landscapes of Iowa and had enjoyed the generosity of colleagues at Iowa State University in sharing their data and their knowledge. We worked from this platform to propose how we would construct and test alternative futures for Corn Belt agricultural watersheds. We conducted this work from 1997 to 2003 with the support of the U.S. Environmental Protection Agency and the National Science Foundation. In Part 2 of this book, particular individuals and institutions are identified for their additional contributions to our efforts.

With insights from Don Boesch, then Director of the Louisiana Universities Marine Consortium, and funding from the National Oceanic and Atmospheric Administration, Nancy Rabalais began documenting the size and dynamics of the Gulf of Mexico dead zone in 1985. Public concern about the growth in the dead zone prompted the federal agencies to begin a thorough assessment of its causes, consequences, and potential remedies in 1997. Under the White House Office of Science and Technology

---

1. U.S. EPA STAR grants program (Water and Watersheds, grant #R-825335-01).

Policy, Don Scavia brought together Rabalais, Don Goolsby, Bill Mitsch, Pat Brezonik, Vic Bierman, Bob Diaz, Andy Solow, and Otto Doering and a team of over 50 scientists to produce the science base for the Mississippi River Basin (MRB) Integrated Assessment (IA).

The IA itself was constructed by a team of scientists from eight federal agencies. Don Pryor of NOAA insisted on relying only on scientific evidence that passed the highest standards, resulting in an effective and influential IA. However, the IA and related science would have gathered dust if it were not for the enlightened leadership of Bob Perciasepe and Chuck Fox, the two EPA assistant administrators who, in turn, co-chaired the interagency, intergovernmental, and intertribal Task Force on Hypoxia in the Gulf of Mexico. Their insistence on an inclusive process, demand for quantitative goals, and drive to produce a practical and balanced product led to the Action Plan that was delivered to the President and the Congress in January 2001, fulfilling a mandate of the Harmful Algal Bloom and Hypoxia Research and Control Act of 1998.

The two broad interdisciplinary projects in the MRB and in Iowa Corn Belt watersheds, conducted independently, came together in conversations between Joan Nassauer and Don Scavia at the School of Natural Resources and Environment at the University of Michigan. Joined by Mary Santelmann, who led the Corn Belt research, we have worked to articulate the relationship between these projects in order to create better choices for the future of Corn Belt landscapes and the health of downstream communities and ecosystems. The USDA Forest Service provided critical support when the two interdisciplinary projects were joined in this book.[2]

Throughout the production of this book, one person has steadied our work with unwavering professionalism, intelligent critique, and deep commitment to the consequences of this work. Jennifer Dowdell, graduate research assistant at the University of Michigan, has been uber copyeditor of the book from its inception. Without her dedicated participation, we undoubtedly would not have brought this work to its finish with such happy satisfaction, and we especially thank her.

Finally, we dedicated this book to Corn Belt farmers because they live most directly with the consequences of agricultural policy. Our research suggests that, when presented with alternatives that pay attention to the environmental and societal—as well as the financial—implications of changing agricultural landscapes, farmers favor the big picture. We hope this book is helpful to policymakers who are looking for ways to view the big picture as well.

*Joan Nassauer, Mary Santelmann, and Don Scavia*

---

2.    USDA Forest Service, MOA #02JV037.

# Contributors

**Mark E. Clark** is an assistant professor in the department of biological sciences at North Dakota State University. His primary interests are in population ecology of vertebrates and ecological modeling.

**Colette U. Coiner** is a GIS/Network Specialist at the Eastern Oregon Agricultural Research Center of Oregon State University. Her expertise is in mapping and modeling the economic effects of rangeland management strategies and their effects on wildlife and the environment. She has published in *Ecological Economics* and *Journal of Dairy Science*, and has contributed to reference manuals for the Integrated Farm Systems and the Dairy Forage Systems Models.

**Robert C. Corry** is an associate professor in the School of Environmental Design & Rural Development, University of Guelph, Ontario. His research focuses on evaluating designed landscape change, and he has published in *Landscape Ecology* and *Landscape & Urban Planning*. He is a co-recipient of the American Society of Landscape Architects' President's Award of Excellence for his contributions to the USDA government interagency publication, *Stream Corridor Restoration*.

**Richard M. Cruse** is a professor of agronomy at Iowa State University. He leads the college's Agricultural Systems Initiative and is director of the Iowa Water Center. His research interests include soil management and cropping systems, tillage, soil erosion, and water quality. He has published in *Agronomy Journal, Soil Science Society of America Journal, Journal of Soil and Water Conservation*, and *Journal of Soil and Tillage Research*.

**Brent J. Danielson** is an associate professor in the department of ecology, evolution, and organismal biology at Iowa State University. He is interested in linkages between behavior, population dynamics, and the interactions between species as they occur and are influenced by complex landscapes. His work ranges from the recovery of hurricane-ravaged populations of endangered species to the quantification of economically important agro-ecological services. He is particularly interested in all things concerning the evolutionary ecology of small mammals.

**Diane M. Debinski** is an associate professor in ecology, evolution, and organismal biology at Iowa State University. Her research focuses on conservation of rare species, understanding species distribution patterns across the landscape, and restoration of functioning prairie ecosystems. She has published in *Ecological Applications, Landscape*

*Ecology, Conservation Biology, Biological Conservation* and *Ecological Restoration.* She has co-authored two field guides for butterflies.

**Otto C. Doering** is professor of agricultural economics at Purdue University, and specializes in economic policy issues affecting agriculture and natural resources. He has served with the USDA as a policy analyst and participant in the process of designing and implementing U.S. farm programs. He was team leader for the national hypoxia assessment's economic analysis and works with various national groups dealing with environmental policy and technology issues.

**Jennifer A. Dowdell** received her Master's degree in landscape architecture at the University of Michigan's School of Natural Resources and Environment. During her graduate work, she was a research assistant in Joan Nassauer's Landscape Ecology Perception and Design Lab. Before beginning her studies, she worked as a policy researcher and advocate with the American Society of Landscape Architects and as editorial assistant manager with *Landscape Architecture Magazine.*

**Joseph M. Eilers** is a professional hydrologist with MaxDepth Aquatics, Inc. in Bend, Oregon. He specializes in data acquisition for river and lake restoration including the use of hydroacoustics, paleolimnology, and water quality modeling. He has published multiple peer reviewed scientific papers in water resources.

**Catherine L. Kling** is a professor of economics at Iowa State University and head of the Resource and Environmental Policy Division of the Center for Agricultural and Rural Development. Her research addresses methods for improving non-market valuation methods and economic incentives for pollution control, especially in relation to non-point source pollution from agriculture. She is a fellow of the American Agricultural Economics Association, vice president and member of the board of the Association of Environmental and Resource Economists, and a member of U.S. EPA's Science Advisory Board.

**David A. Kirk** has managed his own ecological research company (Aquila Applied Ecologists) in Canada since 1989. His main interest is integrating human resource use with conservation of biodiversity through ecologically sustainable land use practices and the planning and spatial design of protected area networks. He has published one previous book chapter and multiple scientific papers.

**Kathryn E. Freemark Lindsay** is a senior manager for Environment Canada and an adjunct research professor and co-director of the Geomatics and Landscape Ecology Research Lab at Carleton University in Ottawa. She has published multiple scientific articles in international peer-reviewed journals, book chapters, and reports/proceedings, including research conducted in Canada, the United States, and the neotropics.

**Greg F. McIsaac** is an associate professor in the department of natural resources and environmental sciences at the University of Illinois at Urbana—Champaign, where he teaches hydrology and ecosystem management and conducts research on the effects of land management practices on the fate and transport of nutrients and sediment.

**William J. Mitsch** is a professor of natural resources and environmental science and director of the Olentangy River Wetland Research Park at Ohio State University. In

2004, he and Sven Erik Jørgensen of the Danish University of Pharmaceutical Sciences were awarded the Stockholm Water Prize for their pioneering development and global dissemination of ecological models of lakes and wetlands, widely applied as effective tools in sustainable water resource management.

**Joan Iverson Nassauer** is professor of natural resources and environment at the University of Michigan. A fellow of the American Society of Landscape Architects and a Distinguished Practitioner of Landscape Ecology, she specializes in the relationship between landscape ecology and public perceptions of ecological design and planning. Her work on design and planning to enhance water quality and biodiversity in agricultural and metropolitan landscapes has appeared in *Conservation Biology, Journal of Soil and Water Conservation, Landscape Ecology, Landscape Journal, Landscape and Urban Planning, Wetlands*, and other journals.

**Stephen Polasky** holds the Fesler-Lampert Chair in Ecological/Environmental Economics at the University of Minnesota. Polasky served as the senior staff economist for environment and resources for the President's Council of Economic Advisers 1998-1999, and currently serves on the EPA's Science Advisory Board's Environmental Economics Advisory Committee and Valuing the Protection of Ecological Systems and Services Committee. His papers have been published in *American Journal of Agricultural Economics, Biological Conservation, Canadian Journal of Economics, Ecological Applications, Journal of Economics Perspectives, Journal of Environmental Economics and Management, International Economic Review, Land Economics, Nature, Proceedings of the National Academy of Sciences, Science*, and other journals.

**Nancy N. Rabalais** is executive director and professor at the Louisiana Universities Marine Consortium. Her research interests include the dynamics of hypoxic environments, interactions of large rivers with the coastal ocean, estuarine and coastal eutrophication, environmental effects of habitat alterations and contaminants, and science policy. She is an author of 3 books and multiple book chapters and peer-reviewed publications, and has received several research and environmental awards for her work on Gulf hypoxia.

**Heather L. Rustigian** works as a GIS specialist and landscape ecologist for the Conservation Biology Institute in Corvallis, OR.

**Mary V. Santelmann** is an ecosystems ecologist and currently serves as director of the water resources graduate program at Oregon State University. Her research interests range from plant species distributions in wetlands to hydrology and biogeochemistry. She has served as principal investigator for interdisciplinary research projects through the EPA Water/Watersheds Program and the U.S. NSF Biocomplexity in the Environment Program. Her publications include papers in *Ecology, Journal of Ecology*, and *Landscape Ecology*.

**Donald Scavia** is professor of natural resources and environment at the University of Michigan, director of the Michigan Sea Grant Program, and director of the Cooperative Institute for Limnology and Ecosystems Research. He specializes in modeling and assessment of human impacts on freshwater and marine ecosystems. He published in journals such as *Science, Limnology and Oceanography*, and *Estuaries and Coasts*, served as associate editor for *Estuaries* and *Frontiers in Ecology and the Environ-*

*ment,* and received the Department of Commerce Gold Medal for Leadership for his role in negotiating the Gulf of Mexico Hypoxia Action Plan.

**Nathan H. Schumaker** is a research ecologist at the U.S. Environmental Protection Agency research laboratory in Corvallis, Oregon. His research interests are in landscape ecology, simulation modeling, and population viability analysis. Schumaker is the author of the PATCH (Program to Assist in Tracking Critical Habitat) model. He served on the State of Oregon, Governor's 4(d) Scientific Review Team, in 1997 and is author or coauthor of several book chapters as well as articles in journals such as *Ecology, Landscape Ecology,* and *Ecological Applications.*

**Jean C. Sifneos** is a faculty research assistant/statistician in the Geosciences Department at Oregon State University. She enjoys using statistics to unravel and explain patterns in environmental data. She has recently published in *Agriculture, and Ecosystems and Environment* and is a co-author of a book chapter on landscape clusters based on fish assemblages and their relationship to existing landscape classifications.

**R. Eugene Turner** is a distinguished research master at Louisiana State University. He is sometimes an oceanographer with his feet firmly in the watershed, and at other times a wetland ecologist mired in shifting coastal soils.

**Kellie B. Vaché** is a hydrologist at the Justus Leibig University in Giessen, Germany. His research has focused predominantly on water and solute movement in large watersheds, with an emphasis on estimation of the cumulative effects of landscape change. He has published papers in various technical journals, including *Geophysical Research Letters* and *Water Resources Research.*

**Denis White** is a geographer at the U.S. Environmental Protection Agency research laboratory in Corvallis, Oregon. His current research areas are integrated environmental modeling and energy systems analysis.

**JunJie Wu** is E. N. Castle Professor of Resource and Rural Economics at Oregon State University. His main research interests include the optimal design of agri-environmental policy, interactions between agricultural production and water quality, land-use patterns and their socioeconomic and environmental consequences, and the rural-urban interface.

# From the Corn Belt to the Gulf

*Societal and Environmental Implications of Alternative Agricultural Futures*

## Chapter 1

# Introduction: Policy Insights from Integrated Assessments and Alternative Futures

## Donald Scavia and Joan Iverson Nassauer

Agriculture in the Mississippi River Basin (MRB) has been identified as the leading cause of depleted oxygen in the "dead zone" of the Gulf of Mexico. The next generation of American agricultural policy could ameliorate this problem along with a host of environmental and societal impacts throughout the MRB. Agricultural policy has powerfully influenced land use and management in the Corn Belt states of the MRB (Figure 1-1), and policymakers are confronted with the question of how changes in these policies can create demonstrable societal benefits both within the Corn Belt and downstream to the Gulf of Mexico.

Negotiations with America's trade partners, along with increasing societal attention to both the costs and environmental effects of agricultural policy, create momentum for change. Emerging technologies and markets for Corn Belt agricultural products, including a surge of interest in ethanol production, open the door further to new possibilities. Pointing to plausible new policy directions and looking at their possible effects over coming decades, this book synthesizes scientific integrated assessments (IAs) at two scales: the entire MRB flowing into the Gulf of Mexico and the local scale of small Corn Belt watersheds where farmers make decisions on the ground. For both IAs, the chapter authors drew on experts in many disciplines to provide quantitative measures of environmental and societal effects of the future landscape changes that agricultural policy could make possible.

The IAs suggest that impressive environmental benefits can be linked to specific changes in Corn Belt farming practices and landscape patterns. At the same time, the IAs address the economic considerations and stakeholder perceptions that will be fundamental to successful change. Both assessments knit together numerous scientific perspectives to give new insight into how policy can simultaneously accomplish many different societal goals (Color Figure 2).

Together, the two link the science on causes of the dead zone with scientific assessments of alternative future scenarios for agriculture in the Corn Belt, and lead to the conclusion that agricultural policy that aims for environmental benefits can improve water quality and relieve flooding along the Mississippi and hypoxia in the Gulf, and

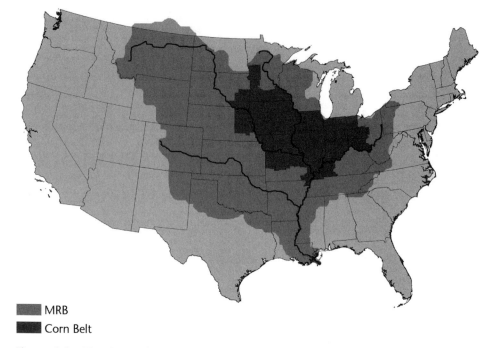

MRB
Corn Belt

**Figure 1-1.** The Corn Belt Region in the Middle and Upper Mississippi Sub-Basins of the Mississippi River Basin (MRB)

*Source:* MRB adapted from U.S. EPA (2001); Corn Belt adapted from USDA ERS (2006).

in doing so, it also can accomplish much more. Policies that address water quality also can help to enhance biodiversity (Color Figures 1 and 3), maintain rural communities, and stabilize farmer incomes. Some new agricultural enterprises that benefit environmental quality (e.g., more extensive perennial cropping), may even help to ameliorate climate change. Furthermore, compared with the agricultural policy of 2000–2006, policies that focus on conservation benefits may be more consistent with international trade mandates.

Closer to home, the landscape qualities resulting from such policies could make Corn Belt landscapes healthier and more attractive places in which to live. This book acknowledges that past agricultural policy has had significant environmental impacts, but it suggests that policy could contribute dramatically to achieving more environmentally beneficial agriculture.

The book is organized in three parts. Part 1 summarizes the environmental and societal drivers that could promote changed agricultural policy. In response to these drivers, Part 2 introduces different policy scenarios and alternative futures for local Corn Belt watersheds and assesses their likely effects. Part 3 summarizes the results of the IAs in both the MRB and the Corn Belt, and describes their convergent implications for agricultural policy.

In Part 1, Chapter 2 reviews the substantial scientific literature on the environmental effects of agriculture in the MRB and its consequences for the Gulf of Mexico dead zone. Chapter 3 describes recent federal agricultural policy as it relates to environmental impacts, rising societal concern for environmental health, and changing expectations for broader societal benefits from agricultural policy.

Part 2 shows how Corn Belt landscapes could change under different future policy scenarios. Chapter 4 describes three very distinct policy-driven scenarios for future Corn Belt agricultural landscapes and shows the alternative futures they could effect for two small Iowa watersheds. These alternative futures for the Iowa watersheds are pictured in the Color Section of the book. The remaining chapters in Part 2 present quantitative assessments that compare the effects of the alternative scenarios on the Iowa study watersheds from the standpoint of economic return (Chapter 5), farmers' perceptions (Chapter 6), water quality (Chapter 7), and plant and animal biodiversity (Chapters 8–13).

Part 3 describes the conclusions drawn from the IAs at each scale—the Corn Belt watersheds (Chapter 14) and the MRB (Chapter 15). Finally, Chapter 16 outlines suggestions for adapting past conservation policies, as well as learning from policy successes elsewhere, to realize agricultural landscapes that more fully achieve environmental and societal benefits, both in local watersheds and across the MRB.

## Two Scales of Integrated Assessment

The IA is the core method of analysis used for this work, and we give an overview of the method in this section. Because this book juxtaposes IAs at two scales (the MRB and second-order Iowa study watersheds), it offers new insights for policy. Each IA used a different approach appropriate to its scale, but the methods and results of the two are highly complementary. The Corn Belt IA was based on development and evaluation of prospective normative scenarios and modeling of concomitant landscape futures for the Iowa study watersheds (Nassauer and Corry 2004). The MRB IA was based on development and evaluation of predictive scenarios (CENR 2000). Both IAs focus on landscape change, defined as spatially explicit changes in land use or land cover and their implications for environmental conditions. Both IAs use science as a means of assessing multiple environmental conditions and share the following common characteristics:

- Assess conditions that are relevant to policy questions.
- Use empirical data that describe past and current landscape conditions to develop descriptions of future landscape conditions.
- Describe policy-relevant characteristics of future landscape conditions.
- Quantitatively compare the function of future conditions on numerous policy-relevant dimensions.
- Integrate multiple assessments to result in a whole picture, allowing comparisons among different future conditions.

### *Mississippi River Basin Integrated Assessment*

In Chapters 2 and 15 we describe results of the IA on the causes and consequences of hypoxia in the northern Gulf of Mexico. This predictive, trend-based IA took the following specific approach (CENR 2000):

1. **Define the policy-relevant question** around which the assessment is to be performed. This was established early in the Harmful Algal Bloom and Hypoxia Research and Control Act (HABHRCA) of 1998, which required this IA with the following statement: "Assess the causes and consequences of hypoxia in the northern Gulf of Mexico."

2. **Document the status and trends** of appropriate environmental, social, and economic conditions related to the issue. This is a relatively value-independent description of current conditions and, to the extent possible, the historical trends in those properties.

3. **Describe the environmental, social, and economic causes and consequences** of those trends. This may include simulation, statistical, and other explanatory models and analyses. These descriptions are fact-based, although they are subject to analysis and interpretation.

4. **Forecast likely future conditions** under a range of policy or management actions, or both. These can be quantitative forecasts from models or other trend-analysis tools, and they are subject to considerable scientific evaluation and interpretation.

5. **Supply technical guidance** for the most cost-effective means of implementing each of the resulting options. These efforts aim to give a menu of available approaches, along with some evaluation of their potential for success and cost-effectiveness, to those who are responsible for their implementation.

6. **Assess the uncertainties** associated with the information generated for the previous steps, and outline key monitoring, research, and modeling needs to improve future assessments in this area.

Integrated assessment at this extensive scale is essential to understanding large-scale impacts, which frame potential options for local action. The MRB IA forces attention to scales well beyond the reach of individual or even state-level action. In doing so, it demonstrates that policies and programs that influence local practices and landscapes also have enormous cumulative effects downstream. By examining the causes and consequences of changes in Gulf of Mexico water quality as a consequence of long-term, large-scale changes in land use and agricultural practices in the MRB, this IA demonstrated significant downstream effects on a continental scale (described in Chapter 2). It also prompted national goals to be established, including protecting the economic and social fabric of Corn Belt agriculture, reducing the size of the hypoxic region in the Gulf of Mexico, and improving local water quality (U.S. EPA 2001). It related the goal of reducing the size of the dead zone to a quantitative nutrient load reduction for the MRB, providing a framework within which regional and local goals, such as those discussed for the Corn Belt IA, can be established and implemented. Evaluating policy options at the scale of the full MRB also allows consideration of impacts on other large-scale phenomena; for example, availability of corn for ethanol production or availability of habitat for species adapting to climate change.

### Corn Belt Futures Integrated Assessment

In Chapters 4 through 14, we describe the results of a normative scenario approach to examining Corn Belt agricultural landscape futures under different possible federal agricultural policies. A normative scenario is not a prediction based on trends; instead, it is a plausible prospect of what might be a desirable outcome. Scenario-based futures are increasingly being used to help decisionmakers think about the outcomes of surprising or desired futures that are not necessarily a continuation of current trends (UNEP 2003; Millennium Assessment 2005). In studies to examine landscape change, the product of a scenario is a future landscape represented by spatially explicit representations of land cover patterns and land management practices. The approach, summarized in the list that follows, is described in more detail in Chapter 4 and in great detail in Nassauer and Corry (2004). Although it employs the same broad principles as the predictive IA used for the MRB, its normative intent requires a different specific method to generate alternative futures. Normative landscape futures were developed in this way:

1. **Define policy-relevant questions for the future of Corn Belt agricultural landscapes.** In the Corn Belt futures project, investigators used a broadly consultative iterative process that began with an e-mail inquiry (see Chapter 4; Figure 4-1) to engage experts from around the nation in identifying relevant issues that could or should affect changes in Corn Belt agricultural landscapes by 2025. These issues ultimately led to defining three alternative scenarios.

2. **Document status and trends with existing data.** To establish the alternative scenarios, researchers used the considerable literature on rural social, cultural, and economic trends in the Midwest to define plausible assumptions for each scenario (Chapter 4; Table 4-1). As Chapter 4 details, to develop concomitant landscape futures for the two Iowa watershed study areas, project investigators analyzed status and trends using both geographic information systems (GIS) data and field data (as described in Chapters 6 and 7).

3. **Develop alternative policy scenarios.** In an intensive field-based workshop (Chapter 4; Figure 4-1), selected experts and stakeholders from a wide range of relevant disciplines developed three policy scenarios. Each scenario was intended to be distinctly different from the others, giving policymakers the flexibility to consider the middle ground among these intentional but plausible "extremes." In addition, related to each scenario, workshop participants developed a rich menu of alternative landscape patterns and practices for the Iowa watersheds. This menu was the foundation for constructing alternative landscape futures for each scenario in each study watershed.

4. **Based on the scenarios, develop explicit land allocation models of plausible alternative future landscapes.** These models used the menu of patterns and practices to describe a set of characteristic land covers for each scenario. Then, location and pattern characteristics of the land covers were operationalized into a set of precise, replicable, decisionmaking rules for landscape change.[1] The rules use existing GIS data to construct GIS models of the alternative future landscapes.

5. **Describe the environmental, social, and economic consequences** of the alternative futures. Part 2 of this book summarizes these quantitative assessments for the Corn Belt IA. Each of the normative alternative future landscapes is a new data set for testing the stakeholder response, economic return, and ecological and hydrological performance of landscapes that could plausibly emerge from new policies. As with the predictive IA conducted for the MRB, these tests include simulations along with statistical and other explanatory models and analyses.

---

1. Detailed decisionmaking rules for the Corn Belt IA can be found at http://www-personal. umich.edu/~nassauer/table_of_contents.html.

6.  **Assess the comparative overall performance** of alternative futures. The futures are a means of comparing the alternative scenarios that they demonstrate, and each is compared against the baseline condition. This allows models and measures with different levels of generality, realism, accuracy, and precision to be compared (Santelmann et al. 2004).

The Corn Belt IA of alternative futures contributes greatly to envisioning aims for policy as they would affect local landscapes. Many of the pattern and practice ideas in the alternative futures could be recombined in different ways to achieve varying policy aims. The futures approach has the advantage of allowing site-specific examination of trade-offs and multiple benefits when landscapes are changed to maintain agricultural productivity and improve water quality and biodiversity. The Corn Belt IA shows that a site-specific integrated approach is necessary to achieve any policy goal in a way that allows consideration of trade-offs and multiple benefits.

The significance of using a landscape futures approach compared with past agricultural policy analysis is discussed in Chapter 16, at the book's conclusion. The authors point out that, because the alternative futures are based on explicit, replicable decisionmaking rules, these rules could be applied and tested more broadly, and their environmental benefits could be more broadly accounted for in dollar terms to more fully assess their policy potential. In Chapter 3 of this book, the authors describe how research has demonstrated that the public is consistently willing to make environmental improvements with costs beyond the current valuation of that environmental improvement.

The Corn Belt scenarios are not intended to be detailed roadmaps for getting to the futures. If there were a roadmap, however, it would undoubtedly show several alternative paths—each presenting different trade-offs in everything from farm income to time frames for adoption to public and political acceptability. The scenarios challenge policymakers to consider alternative courses of adoption in order to reach a desirable future.

## Criteria for Evaluating the Effectiveness of Integrated Assessments

Both the predictive approach of the MRB IA and the normative approach of the Corn Belt IA meet criteria for evaluating the effectiveness of IAs and the transfer of knowledge (Clark and Majone 1985; Mitchell et al. 2006):

- **Adequacy** focuses on technical adequacy—a judgment of whether the assessment uses state-of-the-art methods to ensure the quality of information.
- **Value** refers to the policy relevance of the assessment—whether the assessment question is truly important to policy, whether the spatial and temporal scales of the study match policymakers' needs, and whether the results of the assessment are well communicated.
- **Legitimacy**, specifically "civic legitimacy," involves freedom from bias and allows differing views to be incorporated or addressed.
- **Effectiveness** considers the actual impact of the assessment on the policy decision or debate.

In their review of the hypoxia IA, Rabalais et al. (2002) suggest that its strength comes from being shaped through public participation, quality-controlled with independent expert peer review, and based on high-quality monitoring data. In addition, these authors see the IA as policy relevant, broadly integrative and synthetic, and pre-dictive. More specifically, Scavia and Bricker (2006) reviewed the Gulf hypoxia IA in the context of these criteria. Below, we discuss both IAs in this context.

In the MRB IA, technical adequacy was achieved through peer review—conducted by an independent editorial board—of the six detailed background technical reports and through subsequent publications in the primary literature. Technical adequacy in the Corn Belt IA was achieved through extensive, iterative use of expert panels and expert participation in scenario development and futures review, as well as by ordinary quality assurance procedures in the use of geocoded data.

Value in the MRB IA was ensured because it responded to directives of the 1998 HABHRCA and to the guidance and expectations of the task force that statute created to develop an action plan. Because it was grounded in the considerable scientific litera-ture that describes environmental degradation in the Corn Belt and anticipates chang-ing federal agricultural policies, the policy relevance of the Corn Belt IA was ensured.

Engaging a wide range of stakeholders in the development and review of the MRB IA, as well as formally solicited public comment, facilitated open scientific meetings and public meetings of the task force and ensured legitimacy. The Corn Belt IA's legiti-macy was established differently (described in detail in Nassauer and Corry 2004), by explicitly and intentionally defining scenarios to encompass extreme (but plausible) future conditions, and by offering landscape futures as hypotheses about the future that are described by replicable, explicit models.

The MRB IA was effective because it led to the development of the action plan required by the 1998 HABHRCA (entitled *Action Plan for Reducing, Mitigating, and Con-trolling Hypoxia in the Northern Gulf of Mexico by the Mississippi River/Gulf of Mexico Wa-tershed Nutrient Task Force;* U.S. EPA 2001). This plan, which has been endorsed by eight federal agencies, nine basin states, and two tribes, was delivered to the president and Congress in 2001. Both the MRB IA and the Corn Belt IA continue to influence the way scientists and policymakers think about and discuss policy options, another measure of effectiveness.

## Nested Scales of Analysis Offer New Insight

A unifying theme in this book is that agricultural policy must be crafted and its im-pacts considered across multiple spatial scales. Simply stated, policy must work for a farmer planning a particular field at the same time that it must work as a part of in-ternational trade negotiations. Some of the changes we investigate in this book, such as large-scale restoration and creation of riverine wetlands and fertilizer restrictions, can be meaningfully evaluated only at broad regional scales. In addition, broad-scale analysis is necessary to target locations of local landscape changes, such as changes in agricultural practices. Conversely, the costs and effectiveness of cumulative impacts of alternative futures for small watersheds must be evaluated at the scale of large river basins. The examination of smaller scale alternative future landscapes, though, dem-onstrates that trade-offs to improve water quality and biodiversity while protecting agricultural livelihoods and communities are site specific. From this, it is evident that we must take a site-specific integrated approach to achieve any goal in ways that allow consideration of real trade-offs and multiple benefits.

New agricultural policies must be attentive to their simultaneous effects at local, national, and international scales. To anticipate and use these effects for their greatest public good, it makes sense to explore small watersheds and the context of those small watersheds in the MRB, North America's largest watershed. Do we dare to explore innovative alternatives that may, from a larger spatial and temporal perspective, result in a stronger agricultural economy, healthier rural landscapes, and healthier oceanic ecosystems? We ask those who read this book to look both within and beyond the existing landscapes of the Corn Belt and the MRB and consider not only what is, but also what could be.

# PART 1

## *Environmental and Societal Drivers of Agricultural Landscape Futures*

## Chapter 2

# Corn Belt Landscapes and Hypoxia of the Gulf of Mexico

R. Eugene Turner, Nancy N. Rabalais,
Donald Scavia, and Greg F. McIsaac

L ocal linkages between land use and water quality have cumulative effects within local watersheds, and these effects aggregate up to continental watersheds and their
receiving coastal waters. Although population growth, land use, and climatic events
all have ecological effects, changes attributable to agriculture are the most powerful
links between the Corn Belt region and hypoxia of the Gulf of Mexico. Following
the domination of North America by European culture, and concomitant with urban
growth, a high-intensity agricultural system emphasizing corn production developed
in the midwestern United States. Today, this region is known as the Corn Belt (Chapter
1; Figure 1-1), and current significant water quality problems in the Mississippi River
Basin (MRB) and the Gulf are related to this agricultural landscape change. In the Gulf,
phytoplankton production offshore responds to water quality changes in the Mississippi River that are caused primarily by agriculture. This phytoplankton production
becomes organic material that consumes the oxygen in the bottom layer of the Gulf.
If the oxygen is not replaced as quickly as it is removed, the oxygen concentration
declines, perhaps low enough to become "hypoxic" at less than 2 mg $L^{-1}$. Because mo-
tile organisms will move out of the area when possible and burrowing crabs or worms
(food for fish) can die, this is described in layperson's terms as a "dead zone." This is
the largest dead zone in U.S. coastal waters and in the entire western Atlantic, stretch-
ing across the northern Gulf of Mexico on the Louisiana/Texas continental shelf in
an estimated maximum area (in 2002) about the size of the state of Massachusetts (up
to 5.4 million acres [22,000 $km^2$]; Figure 2-1). Since systematic measurements began
in 1985, the size of the dead zone has increased, and the Gulf's productive diatom–
zooplankton–fish food web, which supports 25 percent of U.S. fish landings, is now
poised to change to one with diminished fisheries (Turner et al. 1998). To minimize,
mitigate, and control this hypoxic zone, a task force of federal, state, and tribal representatives established a goal of reducing the five-year running average of areal extent
of the Gulf of Mexico hypoxic zone to less than 1 million acres (5,000 $km^2$; U.S. EPA

2001; see Table 2-1). This task force recognized the significant role that agriculture in the MRB will need to play to achieve this goal as nitrogen loads to the Gulf are reduced by 30 percent (CENR 2000; U.S. EPA 2001).

Agricultural practices also affect water quality in local watersheds of the Corn Belt. In September 1996, 1,557 fish consumption advisories were issued in the MRB (U.S. EPA 1998). Forty-four percent of the surveyed rivers in 15 MRB states were "impaired" in 2000 (U.S. EPA 2002). In the heart of the Corn Belt in Iowa, the concentration of nitrate in the Des Moines River is often >10 mg L$^{-1}$, which is the statutory maximum limit for drinkable water supplies. Water quality changes in the lower Mississippi River in the 20th century have also been linked to the dead zone in coastal waters (Rabalais et al. 2002).

In this chapter we discuss relationships among agricultural landscape patterns, practices, and water quality in the MRB and the adjacent continental shelf ecosystem of the Gulf. We cite experiments on small watersheds where variables can be controlled, along with anecdotal accounts dating to before the 1930s, to support the conclusion that the release of nutrients stored in the MRB within the pre-European period was large. Various analyses of the variability in water quality among watersheds within the last few decades support the conclusion that changes in land use and population density are directly related to higher nitrogen yields. A paleoreconstruction of continental shelf sediments confirms the significance of these water quality changes

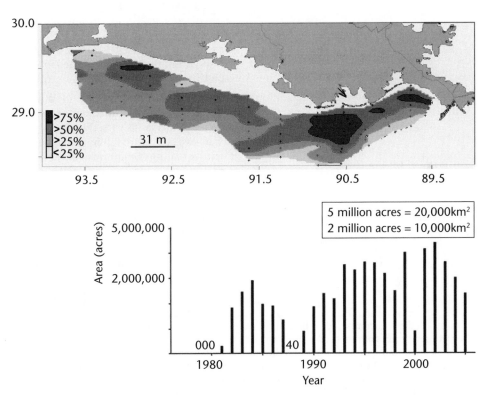

**Figure 2-1.   Top:** Frequency of Midsummer Hypoxia over a 60 to 80 Station Grid from 1985 to 2002. **Bottom:** Variation in the Total Size of Hypoxic Bottom Water in July from 1978 to 2005

*Source*: Modified from Rabalais et al. 2002 and Turner et al. 2006.

to coastal food webs and the development of a coastal zone of low oxygen (the dead zone) in the Gulf.

We also suggest that the most recent influence on nutrient loading, largely intense and widespread farming, has had a more significant effect in the MRB than all previous landscape changes. Linkages among land uses, nitrogen loading to the Gulf, and the size of the hypoxic zone in the Gulf are sufficiently robust to justify support for exist-ing governmental agreements to reduce the size of the hypoxic area and to encourage interest in complementary policies for agricultural conservation.

## Water Quality with Deforestation and Land Drainage for Agriculture

In part because of diseases introduced by the Europeans, the population density of the Native American in the MRB was relatively low in the early 1600s, estimated at no

**Table 2-1.**  Task Force Action Plan Goals for Improvement of Water Quality (U.S. EPA 2001)[a]

| Goal | Description |
| --- | --- |
| **Coastal**[b] | By 2015, subject to the availability of ad-ditional resources, to reduce the five-year running average areal extent of the Gulf of Mexico hypoxic zone to less than 1 million acres (5,000 km²) through implementation of specific, practical, and cost-effective vol-untary actions by all states and tribes and to address all categories of sources and re-movals within the Mississippi–Atchafalaya basin so as to reduce the annual discharge of nitrogen into the Gulf |
| **Within-basin** | To restore and protect the waters of the 31 states and tribal lands with the Mississippi–Atchafalaya basin through implementation of nutrient and sediment reduction actions to protect public health and aquatic life as well as reduce negative impacts of water pollution in the Gulf of Mexico |
| **Quality-of-life** | To improve communities and economic conditions across the Mississippi–Atcha-falaya basin, in particular the agriculture, fisheries, and recreation sectors, through improved public and private land manage-ment and a cooperative, incentive-based approach |

[a]The task force was composed of federal agencies (the Environmental Protection Agency, the Department of Agriculture, the National Oceanic and Atmospheric Administration, the Depart-ment of Interior, the Army Corps of Engineers, the Department of Justice, the Office of Science Technology and Policy, and the Council on Environmental Quality; state governments (Ar-kansas, Illinois, Iowa, Louisiana, Minnesota, Mississippi, Missouri, Tennessee, and Wisconsin); and tribal organizations (the Mississippi Band of Choctaw Indians and the Prairie Island Indian Community).

[b] Task force members agreed to a series of steps to organize around sub-basin strategies to reduce nitrogen loads to the Gulf of Mexico by 30 percent, primarily though voluntary, incentive-based action mediated by education activities.

more than 106,000 or <0.1 person/km$^2$ (Ubelaker 1992). During the 1800s, vegetation cover within the MRB was reduced as European populations and settlement expanded. Population growth in the midwestern states reached 1 to 10 persons/km$^2$ by the 1850s when the population center of the United States crossed the Appalachian Mountains and headed into the MRB in a west–southwest trajectory.

The new settlers brought with them row farming at an intensity previously unknown to these lands, and forests were rapidly cleared for higher-intensity farming. The area being cultivated rose with population growth, but was preceded by removing trees, often by girdling the trunk or burning. Trees were not routinely sold, so we must infer changes in forest area from records of land use. The area of forest in Ohio, for example, went from 54 percent in 1853 to 18 percent in 1883 (Leue 1886). Greeley (1925) documented how the virgin forests of 1850 in the United States were largely remnants by 1920. Humphreys and Abbot (1876) estimated that 15 percent of the MRB was under cultivation or "improvements" by 1860.

When soils are sufficiently disturbed by cultivation, ecological processes that keep nutrients bound up in the soil and organic matter are subdued, and stored nitrogen is released rapidly until the soil's natural fertility is diminished to the point that crop growth is compromised. In the MRB, this was the equivalent of the "slash and burn" agriculture practiced in the tropics, in which crops are planted amidst a shifting mosaic of soils that are newly exploited, in decline, or abandoned to natural rehabilitation. American presidents Washington and Jefferson wrote about the importance of maintaining soil fertility, and the naturalist John Muir observed:

> At first, wheat, corn, and potatoes were the principal crops we raised; wheat especially. But in four or five years the soil was so exhausted that only five or six bushels an acre, even in the better fields, were obtained, although when first plowed twenty and twenty-five were the ordinary yield. More attention was then paid to corn, but without fertilizers the corn crop also became very meager. At last it was discovered that English clover would grow on even the exhausted fields, and that when plowed under and planted with corn, or even wheat, wonderful crops were raised. (Muir 1965, Chapter VI: "The Ploughboy," *85*)

Gray's review (1933) of southern agriculture up to 1860 characterized the attitude and consequence of "soil exhaustion" as an expected consequence of a way of farming:

> Planters bought land as they might buy a wagon—with the expectation of wearing it out . . . as the wave of migration passed like a devastating scrooge [sic]. Especially in the rolling piedmont lands the planting of corn and cotton in hill and drill hastened erosion, leaving the hillsides gullied and bare. (Gray 1933, *446*)

The application of technological inventions in the 1800s introduced changes to farming practices. The widespread use of the steel moldboard plow, threshing machine, mower, and reaper began between 1825 and 1850. The first agricultural journals were issued, such as *The American Farmer* (Baltimore, 1819); *The Plow Boy* (Albany, 1819); *The New England Farmer* (Boston, 1822); *The New York Farmer* (New York, 1827); and *The Southern Agriculturist* (Charleston, 1828). Land management practices evolved in the direction of less land clearing and more intensive use and replacement of native

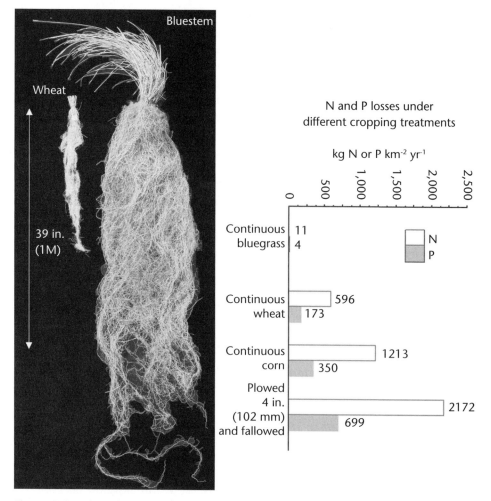

**Figure 2-2.** The Belowground Vertical Profile of the Roots in Wheat (*Triticum aestivum*) and Native Big Bluestem (*Andropogon gerardi*) Prairie Vegetation.

*Notes*: Experiments were conducted at the Missouri Agricultural Experiment Station (Miller and Krusekoff 1932). Average nutrient losses were measured from May 1, 1926 to May 1, 1928. Losses were measured in runoff from experimental plots using different cover crops and farming practices. Fertilizers were not applied in these experiments.

*Source*: Turner and Rabalais 2003.

Photo courtesy of Karena Schmidt, U.S. Department of Agriculture (USDA) Agricultural Research Service (ARS)

deep-rooted cover (perennials) with shallow-rooted annual crops. These newer approaches to soil management led to high nutrient losses relative to those from continuously vegetated land. A two-year study from one of the new agriculture experiment stations in the early part of the 1900s documented that when unplowed prairie was converted to continuous wheat or corn, nitrogen (N) and phosphorus (P) losses were approximately 43 to 110 times more than under perennial grass (Figure 2-2).

In 1835, John Johnston (1791–1880) became the first person in the United States to lay drain tile on his farm in Ithaca, New York, and drainage improved crop yields on marginal lands. It was not until the Swamp Land acts (see http://www.npwrc.usgs.gov/

resource/wetlands/uswetlan/century.htm) passed and the railroad companies (which owned vast tracts of land) expanded, however, that Corn Belt farmland began to be drained. As highly productive row crop farming expanded into formerly waterlogged soil, drainage increased soil nitrogen losses (McIsaac and Hu 2004). The use of tile drainage appears to have increased since the 1990s.

## Water Quality after Widespread Use of Fertilizers

The first patent for a chemical fertilizer was issued in 1849 to the Chappell brothers in Baltimore, and subsequent phosphate fertilizer production was centered near Charleston, South Carolina, in the late 1800s. This early local use of fertilizer was dwarfed by the large-scale production and use of nitrogen fertilizers that began after World War II. Fertilizer nitrogen is now the dominant nitrogen source in the MRB (Figure 2-3).

Nitrogen losses from soils are increased when soils are ditched and drained. The majority of dissolved nitrogen lost in a drained field is in the form of nitrate, which moves in shallow, subsurface flow or as deeper groundwater rather than in overland flow (Lowrance 1992). Several studies have documented a strong and proportional relationship between nitrogen applications to surface soils and the nitrogen content in soil pore water and groundwater (Baker and Laflen 1983; Hallberg 1989; Keeney 1986). Nitrate accumulation in soil water of experimental plots in Virginia and Nebraska, for example, increased as nitrogen fertilizer application rates increased (Hahne et al. 1977; Schepers et al. 1991). Tile drainage, on the other hand, promotes nitrate leaching losses to groundwater, which eventually makes its way to surface waters (Hallberg 1989; McIsaac and Hu 2004).

Increased fertilizer use was accompanied by a rise in the nitrate content of the Mississippi River at the "end of the pipe" near New Orleans, Louisiana (Figure 2-4). The loading of nitrogen varies monthly, and the annual average load from the Mississippi River is now two to three times higher than it was in the 1960s. Water quality measurements at New Orleans and upstream at St. Francisville, Louisiana, taken in 1905–1906, 1933–1934, and the early 1950s suggest that the significant increases in nitrate were

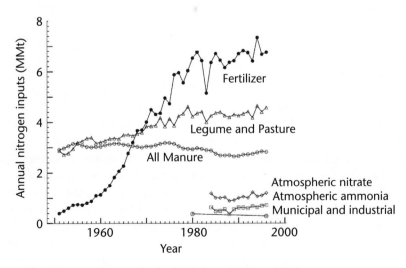

**Figure 2-3.** The Nitrogen Sources in the MRB from 1985 to 1998
*Source*: From Goolsby et al. 1999.

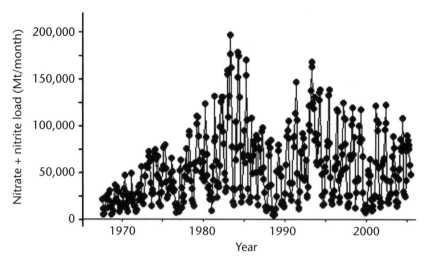

**Figure 2-4.**   The Monthly Nitrate + Nitrite Load at St. Francisville, Louisiana, from July 1967 to June 2005

*Source*: Modified from Turner et al. 2005.

seen during the period when fertilizer applications in the watershed increased (Turner and Rabalais 1991).

The sequence of colonization, land clearing, agricultural expansion, soil erosion, and increased nitrogen loading is summarized in Figure 2-5 in panels (a), (b), and (c). The (d) panel is a record of the marine diatoms deposited in the seabed. Diatoms are the major algal prey of the phytoplankton that are near the base of the fish coastal food web. Diatoms incorporate silica in their cell walls. The diatoms produced in the water column are indirectly estimated by examining the sediments for their remnants by measuring the biologically bound silica (BSi). The BSi peaks and declines are coincidental with the land use changes resulting from land clearing, the expansion of agriculture, and land drainage efforts within the MRB, and are modulated by the natural restoration of abandoned land. Note that the percentage of BSi in sediments accumulating offshore is higher now compared to the peak in the mid-1800s. This recent rise in BSi concentration is clearly related to the increased nitrogen loading from the river that occurred as a direct consequence of the dramatic rise in fertilizer application after World War II (Figure 2-5[d]).

Present-day delivery of water, sediments, and nutrients from the major sub-basins of the Mississippi differs among basins, and the amount of water discharged does not necessarily correlate with the amount of sediment or nutrients discharged (Figure 2-6). From 1973 to 1994, the Ohio watershed delivered an average 38 percent of the fresh water discharged into the Gulf of Mexico. The Upper Mississippi discharged 19 percent of the total, followed by the Missouri (13 percent), the Lower Mississippi (13 percent), the Red and Ouachita (11 percent), and the Arkansas (11 percent). The Missouri watershed has 42 percent of the land surface, compared to almost equal percentages by the Ohio, Upper Mississippi, and Arkansas watersheds (16, 15, and 13 percent, respectively). The remaining 14 percent of land area is almost equally divided between the Red and Ouachita watershed and the Lower Mississippi watershed. The Missouri is the largest watershed among the six, by at least 2.5-fold, but the water yield in the Ohio watershed is three times higher than the others (Turner et al. 2000).

Sediment yields from 1973 to 1994 were highest in the Lower Mississippi watershed, which is about two times higher than those of any of the other watersheds, and four times higher than the MRB as a whole. The Arkansas watershed has the lowest yield (20 percent of the average for the MRB). The proportional amount of suspended sediment supply is dominated by the Missouri watershed, which yields 42 percent of the total supply, and is two times more than that from the Lower Mississippi watershed.

The highest nitrate yield is from within the Upper and Middle Mississippi watershed at the center of the Corn Belt. The highest total phosphorus (TP) and total nitrogen (TN) yields are from the lower Mississippi River, which also has the highest suspended sediment yields. The highest silicate and total organic carbon loading is from the Ohio–Tennessee watershed. The ranges of yields vary from 14 to 194 kg km$^{-2}$ yr$^{-1}$ (79.93 to 1,107.74 lb mi$^{-2}$ yr$^{-1}$) for TP; from 26 to 902 kg km$^{-2}$ yr$^{-1}$ (148.46 to 5,150.42 lb mi$^{-2}$ yr$^{-1}$ ) for nitrate; from 141 to 1,120 kg km$^{-2}$ yr$^{-1}$ (805.11 to 6,395.2 lb mi$^{-2}$ yr$^{-1}$) for TN; from 262 to 1,074 kg km$^{-2}$ yr$^{-1}$ (1,496.02 to 6,132.54 lb mi$^{-2}$ yr$^{-1}$) for silicate; and from 783 to 2,513 kg km$^{-2}$ yr$^{-1}$ (4,470.93 to 14,349.23 lb mi$^{-2}$ yr$^{-1}$) for total organic carbon (TOC). Four of the lowest values (including suspended sediments) are for the Arkansas watershed.

Relative loadings among watersheds are not the same as the relative yields because land mass size varies among watersheds. From 1974 to 1994, the upper three watersheds had about equal loadings of TOC (ranging from 23 to 32 percent of the total). The loadings of TP and silicate are about equally divided among the Upper and Middle Mississippi, Lower Mississippi, Ohio, and Missouri watersheds. The TN and nitrate loadings from the Red and Arkansas watersheds are relatively small compared with the others (<7 percent each). Important for its effects on downstream hypoxia in the Gulf, the dominant watershed in terms of TN and nitrate loading is the Upper and Middle Mississippi watershed (35 and 45 percent, respectively), followed by the Ohio watershed (28 and 30 percent, respectively). These are the watersheds of the Corn Belt states. Scavia et al. explore the policy implications of nitrate loading coming primarily from the Corn Belt in Chapter 15 of this book.

Compared to the beginning of the 20th century, the nitrate load in recent decades is higher by 150 percent for the MRB as a whole. Over this time, proportions of nitrate loading from each watershed have been redistributed toward a higher yield from the Upper and Middle Mississippi watershed, and the proportional yield from the Ohio watershed has declined. The proportional amount from the Missouri watershed has remained about the same. The percentage from the Red and Arkansas watersheds remained less than 5 percent. From these analyses, we can clearly conclude that the Upper and Middle Mississippi River watershed, including the Corn Belt, is a large contributor to the TN yield flowing downstream to cause the dead zone in the Gulf.

## Agricultural Practices and Water Quality Quantitative Models

Various models have been developed to test how tightly coupled MRB nitrogen loading is to the area of hypoxia (Table 2-2). These models are useful for evaluating progress toward reaching the goal of the action plan and for refining the quantification of the relationships between land use in the MRB and the size of the hypoxic zone. For example, Scavia et al. (2003, 2004) suggest that the required nitrogen load reduction should be as much as 35 to 45 percent, as opposed to the 30 percent called for in the action plan (U.S. EPA 2001). Some models are able to accurately predict the size of the hypoxic zone several months before the annual July survey cruise (Figure 2-7). A lesson

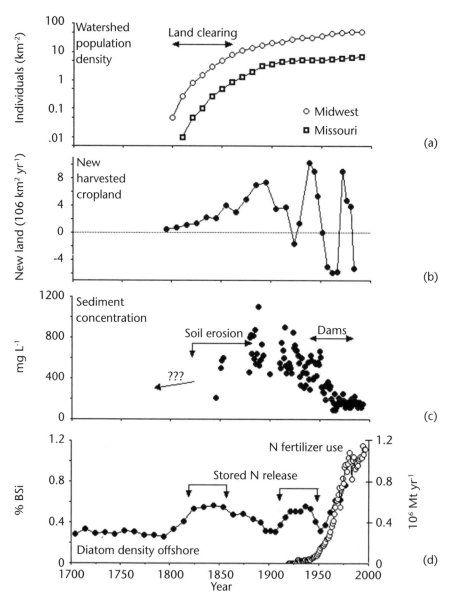

**Figure 2-5.**  A Summary Interpretation of the Relationships among Anthropogenic Activities[a] in the MRB and Coastal Phytoplankton Production

*Notes:*

(a) Population density in the MRB watershed: the Midwest (Ohio, Indiana, Illinois, Iowa, Michigan, Wisconsin, and Minnesota) and the Missouri watershed (Montana, Wyoming, North Dakota, South Dakota, Nebraska, and Missouri). Population data are from the 2000 U.S. Census, and in most cases, exclude Native Americans and slaves before 1850.

(b) The new area of harvested cropland and land under drainage added each year in the MRB. The rise in harvested cropland after World War II is probably taking place on the previously farmed land.

(c) The annual average suspended sediment concentration (mg L$^{-1}$) at New Orleans, Louisiana.

(d) The percentage of BSi in sediments from dated sediment cores collected near the mouth of the Mississippi River (from Turner and Rabalais 1994). The BSi percentage is an indicator of BSi found in diatom remains (dry weight basis). Also included in the bottom panel is the annual N flux in the Mississippi River from 1940 to 1987.

[a] Population growth, land conversion to agriculture, and fertilizer use.

*Source:* Turner and Rabalais 2003.

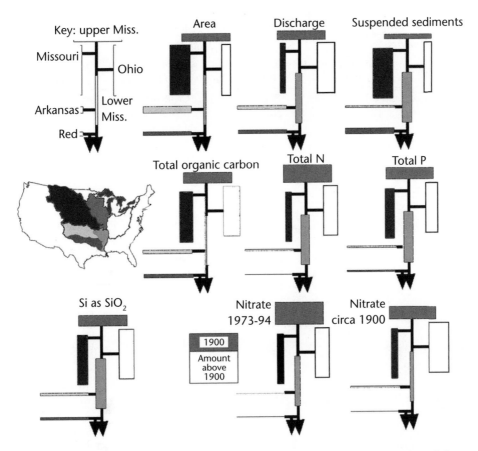

**Figure 2-6.** The Proportional Amount of Seven Factors[a] from 1974 to 1994 (Mass) from Each of Six Watersheds in the Mississippi River Watershed

*Notes*: The estimated proportional amount of nitrate loading (mass) for circa 1900 is also shown. The nitrate loading circa 1900 for the whole watershed was about 38 percent of what it is today. The two arrows leaving the watershed are for the Mississippi River (right) and the Atchafalaya River (left), which joins the Red River with about 30 percent of the Mississippi River flow.

[a]Watershed area, water discharge, suspended sediment loading, TOC, TN, TP, and nitrate, expressed as a percentage of the total

*Source*: From Turner and Rabalais 2004, with kind permission of Springer Science and Business Media.

learned from these models is that nitrate loading, by itself, is a significant correlate of the size of the hypoxic zone (in late July, when the best data set is available and, by implication, throughout the year). The appearance of a significantly large hypoxic zone before the 1970s is unlikely, which would be consistent with the relationships implied by the coherence of land use change and increased nitrogen loading discussed earlier in this chapter. The models also indicate that the carbon accumulating in the sediments from the higher marine diatom production (stimulated by higher nitrogen loading to the coastal zone) will continue to deplete oxygen from the water column even if nitrate loading were greatly reduced, thus acting as an "ecosystem memory" that will lengthen recovery times.

Several quantitative investigations have demonstrated the relationship between land use, nitrogen, phosphorous, and water quality. Jordan et al. (1997) showed that nitrate yields increased and the silicate:nitrogen yield decreased as the watershed area

**Table 2-2.**   Models Used to Forecast the Size of the Mid-Summer Hypoxic Zone

| Model elements | Models | | | |
|---|---|---|---|---|
| | *Bierman et al. (1994)* | *Justic et al. (1996, 2003a)* | *Scavia et al. (2003, 2004)* | *Turner et al. (2005)* |
| Type | 21-segment, 3-D grid | Two vertical boxes | Continuous linear coverage, adaptation of river model | Statistical |
| Area | Shelfwide | One location in area of most frequent hypoxia | Shelfwide | Shelfwide |
| Calibration years | 1985, 1988, 1990 | 1985–1993 | 1985–2002 | 1978–2004 |
| Key variables | Nutrient load, water circulation, temperature, solar radiation, light attenuation | Discharge, nitrate concentration and load, temperature, wind | Nitrogen load; year-specific advection term | Nitrate load, year |
| Sediment oxygen demand included? | Yes | No | No | Indirectly |
| Lag period for nitrogen load | Not applicable | One month | May–June average | Two months |
| Relative N versus P response of reduction or effect | N > P | Not done | Not done | N > P |

as cropland expanded for 27 watersheds of Chesapeake Bay. Smart et al. (1985) studied watersheds in the Missouri Ozarks in the summer of 1979 and found that the silicate: nitrate ratio and the nitrogen content increased as the land in pasture expanded (apparently there were few row crops in that area). They concluded that the stream nutrient concentrations were more strongly related to land use than to bedrock geology. The simple multiple regression equation used by Smart et al. (1985) explained 43 percent of the variation in TP, using watershed size and the percentage of land as urban area. Similarly, the equation explained 80 percent of the variation in TN, using the percentage of land in pasture and urban area. Perkins et al. (1998) showed similar results for all four major types of Missouri watersheds, as did Jones et al. (1976) for 34 watersheds in northwestern Iowa (a three-year data set). The elemental ratios in the water are subject to the type of land use, as well. For example, Arbuckle and Downing (2001) showed that water flowing from 113 Iowa landscapes dominated by row crops had high N:P ratios (molar ratios >100), whereas landscapes dominated by pasture lands had low ratios (molar ratios about 16:1). One interesting pattern in the Jones et al. (1976) data set is that the nitrogen yield dropped rapidly with a small increase in land as marsh, and continued to fall to less than 1 kg ha$^{-1}$ (0.89 lb acre$^{-1}$) when the watershed was 15 percent marsh. This finding suggests that strategic restoration of wetlands throughout

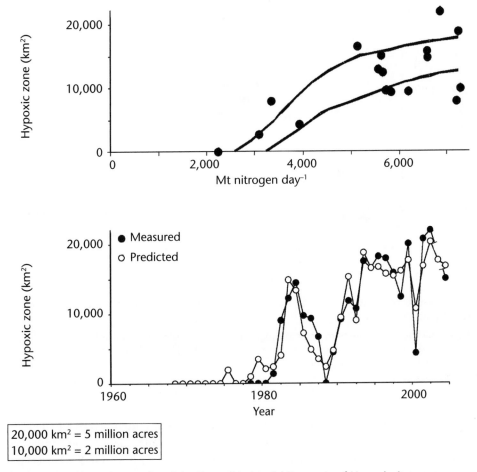

20,000 km² = 5 million acres
10,000 km² = 2 million acres

**Figure 2-7.** **Top.** The Results of the Ensemble Model Forecasts of Hypoxic Area as a Function of May–June TN Load. **Bottom.** The Results of a Final Model Predicting the Size of the Hypoxic Zone from 1968 to 2004

*Notes:* Symbols in the top panel represent data from 1985 to 2002. The curves are the first and third quartiles from Monte Carlo simulations. Model forecasts are described in Scavia et al. (2003). The bottom panel includes estimates over the entire shelf. The equation is Y (km²) = -1337953.4 + 672.1589 * Year + 0.0998 * (may flux as $NO_3$+2). The hindcast values plotted as zero prior to 1978 are negative values in the model.

*Source:* Modified from Turner et al. 2005.

the upper watershed may be a useful water quality restoration approach, especially where the nitrate concentration is high (Turner 2005).

Variations among sub-basins of the MRB have also been quantitatively related to land use. Nitrogen yields from the MRB and coastal watersheds are described quite well by the percentage of land in cropland and population density (Figure 2-8), a representative indicator of nitrogen transformation and sources in the landscape. Howarth et al. (1996) and Peierls et al. (1991), for example, described a strong relationship between population density and nitrogen yields for many of the large river watersheds draining into the north Atlantic. A multiple regression equation, predicting nitrogen yield on the basis of the land in crops and population density, explained 78 to 83 percent of the variation for the three individual data sets, and 60 percent of the variation for all data

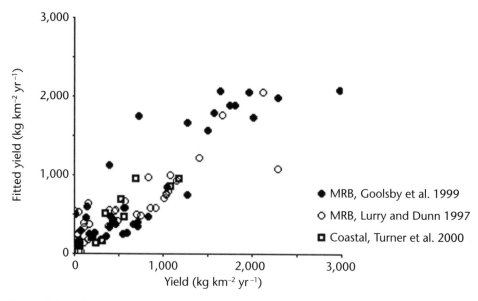

**Figure 2-8.**    The Actual and Estimated Yield of TN for Watersheds within the MRB and U.S. Coastal Systems

*Notes*: The fitted values are from a multiple regression equation with two independent variables: population density and the percentage of land as harvested cropland. MRB data from two sources; coastal system data from Turner et al. (2000).

*Source*: From Turner and Rabalais 2004, with kind permission of Springer Science and Business Media.

sets combined (Turner and Rabalais 2004). The results are strikingly robust and demonstrate how both population density and land use affect water quality. The annual per capita yield in these regression equations ranges from 5 to 21 lb (2.3 to 9.7 kg) TN per capita, with the coastal data sets having a lower per capita yield (6 lb [2.6 kg] TN per capita compared to 12 to 21 lb [5.3 to 9.7 kg] TN per capita for the MRB).

Regional mass balances of reactive nitrogen inputs and outputs provide a quantitative assessment of the sources of reactive nitrogen and can suggest the approximate scale and scope of remediation alternatives. Such mass balances have shown that average annual riverine N flux is highly correlated with average annual net anthropogenic nitrogen input (NANI) to the drainage basin (Howarth et al. 1996; Boyer et al. 2002). NANI is the sum of fertilizer, biological fixation associated with crops, and atmospheric deposition of oxides of nitrogen minus the N exported in food and feed. The values of inputs and outputs can be estimated from annual agricultural and census statistics and data from the National Atmospheric Deposition Program as described by Goolsby et al. (1999), Goolsby and Battaglin (2001), Boyer et al. (2002), and McIsaac et al. (2002).

In many temperate regions, including the MRB, riverine loads of N to coastal waters are an average of 25 percent of the regional NANI. The other 75 percent may be converted to gaseous forms of N or stored in soils or groundwater temporarily or permanently (VanBreemen et al. 2002). There is, however, some variation across drainage basins and over time in the percentage of NANI delivered to coastal waters, which appears to be related to variation in soils, wetlands, and hydrologic efficiency of transporting unutilized N from the land to surface waters.

In Illinois watersheds with extensive artificial subsurface (tile) drainage, McIsaac and Hu (2004) reported that annual average riverine N fluxes were equal to regional

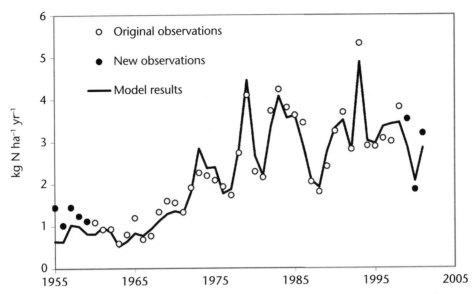

**Figure 2-9.**   Observed Nitrate Nitrogen Flux in the Mississippi River at St. Francisville[a]

*Notes:* Values prior to 1960 and after 1998 (new observations, solid circles) were not included in the original model development because the necessary crop and livestock production data to calculate NANI were not readily available. Subsequent compilation of historical data and collection of new data have allowed calculation of these "new" values.

[a] Open circles and solid circles show observed flux.

NANI. This suggests that denitrification, for which NANI does not explicitly account, may be negligible in tile-drained watersheds, or that depletion of soil organic N is contributing to denitrification and riverine N flux. Further research is needed to refine our understanding of the transport and fate of reactive N in these and other settings. Nevertheless, the correlation between high NANI and high riverine N fluxes is observed across wide geographic regions, and tile drainage appears to increase the transport of NANI from land to surface waters.

In addition to variation in NANI, the quantity of riverine N delivered to surface waters and coastal waters varies annually and seasonally depending on the quantity and quality of precipitation and drainage waters. Nitrate and nitrite N are highly mobile, biologically available forms of N that account for practically all of the increased N flux in the Mississippi River to the Gulf of Mexico in the 20th century (Goolsby et al. 1999). Nitrite concentrations are usually much smaller than nitrate and often considered a negligible influence. Nitrate concentrations and flux tend to increase with river flows.

McIsaac et al. (2001, 2002) developed a regression model that accounted for 95 percent of the variation on annual riverine nitrate plus nitrite N flux in the lower Mississippi River at St. Francisville for the period from 1960 to 1998 (Figure 2-9). The model suggests that relatively small changes in the long-term average NANI would lead to relatively large changes in riverine nitrate flux. It was estimated, for instance, that a 14.2 percent reduction in NANI fertilizer N input would reduce riverine nitrate flux by 34 percent.

Additionally, the model suggests that changes in NANI influence riverine nitrate flux for the succeeding two to nine years, with NANI during the previous two to five years having a greater impact than NANI from the previous six to nine years. This lag time may be caused by hydraulic lags related to flow paths, the cycling of nitrogen through soil organic matter, or both. Additionally, a portion of the lag may result from hydrologic persistence. Large values of NANI coincide with droughts, which leave un-utilized fertilizer in corn and wheat fields at the end of the growing season, and also leave soils and reservoirs depleted of water. Following droughts, nitrate concentrations in drainage waters tend to be larger than normal, but before stream flow returns to normal levels, soil and other water reservoirs need to be refilled. Recharging these reserves also provides opportunities for temporary storage and denitrification of the nitrate carried in the drainage waters. Changes in nitrate flux in the Mississippi River should be interpreted in light of the changes in NANI and stream flow over the previous ten years.

NANI increased from 1960 to 1980 as N fertilizer use increased (Figure 2-10). In the 1980s, NANI was highly variable, in part because of major droughts in 1980, 1983, and 1988. Since 1988, NANI has been generally declining, in large part because crop yields have increased faster than N fertilizer input (Figure 2-11). The amount of fertilizer N applied per bushel of corn harvested has declined since about 1990. In the 1970s, fertilizer was relatively inexpensive compared to the value of the additional corn it produced. Consequently, application rates generally exceeded economic optimum use. Fertilizer use efficiency has improved as the price of fertilizer has increased and technology and information about using fertilizer optimally and economically have improved. The 1997–2001 value of NANI (15.0 kg N ha–1 yr–1 (13.35 lb N acre–1 yr–1) is 10 percent less than the 1976–1996 average value (16.7 kg N ha–1 yr–1 (14.86 lb N acre–1 yr–1). Between 1995 and 2004, weather conditions did not reduce yields over large areas within the basin in any growing season.

Water flux is a major driver of nitrate flux, and the effects of the reduced NANI on nitrate flux will be clearer during periods of above-average water flow. The average annual flux of nitrate to the Gulf of Mexico from 1999 to 2004, however, was approximately 2.4 kg N ha–1 yr–1(2.14 lb N acre–1 yr–1), which is 20 percent lower than the average value from 1979 to 1998 (3.2 kg N ha–1 yr–1(2.85 lb N acre–1 yr–1), and water flux from 1999 to 2004 was about 13 percent lower than the average from 1978 to 1998.

This NANI model (McIsaac et al. [2001, 2002]) is subject to errors because it does not explicitly address several factors that are known to influence the quantity of nitrate losses from agricultural fields: (1) tile drainage (which appears to be increasing rapidly in recent years); (2) timing of fertilizer application (see Chapter 3 in this volume and Randall and Vetsch [2005]); and (3) the timing of excess rainfall and drainage. Additionally, this model lumps all nitrogen into a single category, but in actuality, the different pools of nitrogen have very different pathways and delivery rates. Fertilizer nitrogen cycles through the crop and soil system, while a significant portion of the nitrogen in human waste in treated municipal wastewater is discharged directly into the surface water system. Because of these limitations and confounding factors (e.g., increased tile drainage), the recent reductions in NANI suggested by the model may not lead to the predicted large reductions in riverine nitrate flux. Additionally, reductions in NANI depend, to some degree, on favorable weather conditions for crop production, which are unlikely to continue indefinitely.

For this reason, changes in agricultural landscapes and practices to reduce nutrient flux and protect ecosystems of the Gulf should be multifaceted to address the different sources and variations in flux. Continued refinement of nutrient use efficiency in the existing crop production systems is possible, although it will be difficult or costly

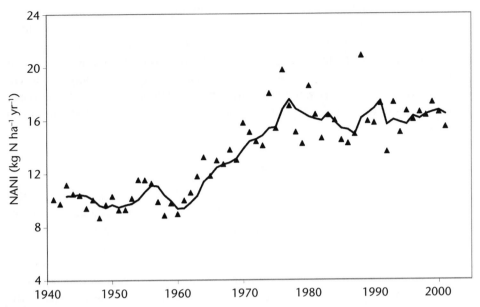

**Figure 2-10.** Annual NANI to the MRB[a] and Five-Year Moving Average Value[b]

[a] Solid triangles, values estimated from available data sources

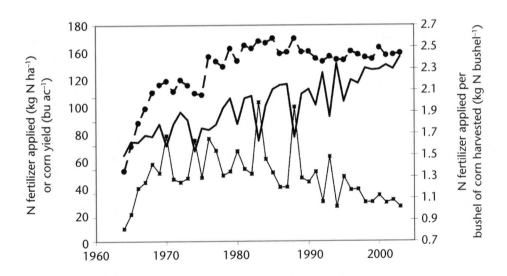

**Figure 2-11.** Area-Weighted Average Commercial N Fertilizer Application to Corn in the Major Corn-Producing States[a]

[a] Iowa, Illinois, Indiana, Minnesota, and Nebraska

*Note*: Top line is fertilizer applied. Center line is corn yield. Bottom line is fertilizer applied per bushel of corn.

*Source*: See http://www.ers.usda.gov/Data?FertilizerUse/ and http://www/nass.usda.gov/index.asp.

to achieve low nitrate losses from corn–soybean production on tile-drained cropland (Jaynes et al. 2004; Karlen et al. 2005). Cropping practices and systems that result in less nutrient losses than the conventional corn–soybean system, like those described in the alternative futures, in Part 2 of this book, may be needed. Cropping systems that utilize perennial vegetation for livestock feed, biomass energy, or carbon or nutrient sequestration tend to reduce water and nutrient losses compared to the corn–soybean system that currently dominates the Corn Belt (Randall et al. 1997). Restoration and protection of wetlands also can reduce nitrate fluxes.

## Conclusions

The scale of water quality changes to the MRB over the last 200 years is substantial, and societal benefits from the agricultural landscape of today are great and important to many. Many are, understandably, quite satisfied with the status quo, which produces food and income. The enormous MRB, however, has not been in anything like an equilibrium system for the last 200 years, and historical precedence suggests that additional changes are forthcoming. In many communities throughout the Midwest, nitrate levels in drinking water exceed national standards, and excess nitrogen is the leading cause of the largest coastal hypoxic zone in the western North Atlantic. In addition, eutrophication of inland and coastal waters is widely acknowledged as a social and environmental concern (Rabalais 2002). Policy choices affecting flood protection and agriculture join with other legitimate societal concerns about diminishing natural resource quality and quantity. Effective action in the MRB could also shape policy agendas involving international trade, global climate change, and competing demands for federal expenditures.

One point seems certain: It took decades for the agricultural landscape to reach its current state, which suggests that it will take decades of consistent policy implementation for water quality rehabilitation to succeed. Understanding and managing time lags in soil nutrient pools and turnover is a key factor in water quality management. Three examples offer perspective. First, the unfertilized fallow soil at the Rothamstead (United Kingdom) experimental soil plots continues to leak significant amounts of nitrogen after 40 years (Addiscott 1988). Second, in Sweden, nitrate leaching from a grain field continued almost unabated for the first 13 years after fertilizer use was discontinued (Löfgren et al. 1999). Third, data on river water quality following the collapse (circa 1990) of agricultural collectivization in the former Soviet republics of Estonia, Latvia, and Lithuania show that, although fertilizer application fell dramatically, the concentration of inorganic phosphate and nitrogen was the same in 1994 as in 1987 (Löfgren et al. 1999). The mineralization of the huge soil nitrogen pool in these Baltic states is 50 to 200 kg N ha$^{-1}$ (44.5 to 178 lb N acre$^{-1}$) compared to 64 to 93 kg N ha$^{-1}$ (56.96 to 82.77 lb N acre$^{-1}$) of fertilizer applied.

To respond to these complex realities, today's soil and land management practices must be much broader than those reflecting the commonly held view expressed in a popular soil textbook from a half-century ago: "After all, our primary aim in soil management is to seek the highest yields we can maintain consistent with greatest profit" (Thompson 1957, *363*).

As the future scenarios described in the following chapters demonstrate, our view of profit can be enlarged to include the environmental and societal costs as well as the benefits of production agriculture. Agricultural policy can do more than support productive agriculture. It can enhance biodiversity, contribute to healthy rural communities, and protect and improve soil and water quality over decades rather than only a few years. How do we achieve these diverse and sometimes competing goals? The chapters that follow suggest that innovation is key—just as it was key to introducing tile drainage and widespread fertilizer use to the Corn Belt in the past century. New policies and practices should build on the successes of the past, and at the same time invite innovation to change the Corn Belt and the Gulf in response to our growing understanding of ecosystems and societal health.

## Chapter 3

# Changing Societal Expectations for Environmental Benefits from Agricultural Policy

## Joan Iverson Nassauer and Catherine L. Kling

Agricultural policy has powerfully influenced land use and land management practices in the Corn Belt, and the environmental consequences extend throughout the region, down the Mississippi River, and all the way to the Gulf of Mexico. Three-quarters of the area of the Corn Belt states—Illinois, Indiana, Iowa, Missouri, Minnesota, Ohio, and Wisconsin—is in farmland. This land produces 25 percent of the nation's agricultural income, and its farmers receive about one-third of all federal agricultural support payments (USDA NASS 2002). As policy changes, the agricultural landscape of the Corn Belt will change.

Public concerns about environmental health and the federal budget, along with rising international attention to the trade effects of agricultural subsidies, have led to new expectations for agricultural policy. In response, American policymakers can plan for Corn Belt agricultural practices and landscapes to change in ways that create demonstrable societal benefits. In this chapter, we examine some environmental effects of past federal agricultural policy. We also consider how changing public expectations as well as new international trade mandates could influence future policy to produce greater societal benefits, including clean, adequate water supplies; healthy lakes and rivers; enhanced biodiversity; more attractive rural landscapes; and more sustainable rural communities.

### Environmental Conditions and Public Perceptions

The signs of environmental degradation associated with American agricultural landscapes are startling and unmistakable (Figure 3-1 and Table 3-1). Drawdowns of major continental aquifers for agricultural irrigation; degraded aquatic habitats and compromised water quality; pollution of drinking water by nitrates leaching from farmland; massive flood destruction of cities, shipping, and industry; and elimination of all but the most fragmented remnants of native ecosystems in the heart of the Corn Belt—all must be attributed in part to past agricultural practices, practices that could be transformed by future policy (Runge 1996; Babcock 2001; Claassen et al. 2001; Tilman et

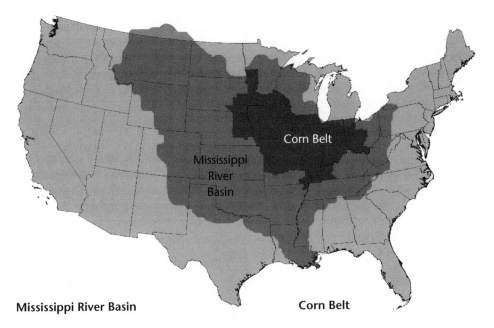

**Mississippi River Basin**

- Ninety-five percent of the prairie, savanna, wetlands, and woodland converted to agricultural lands since permanent settlement began

- Mississippi River carries more nitrate than any other U.S. river

- Nitrogen loading 50 times higher than presettlement rates

**Corn Belt**

- Global flyway for 60 percent of North American bird species

- Habitat for 300 endangered species

- One-third less carbon content in soils than during colonization

**Figure 3-1.** Some Notable Environmental Effects of Agriculture in the Corn Belt States

al. 2001; Gagnon et al. 2004; USDA 2006). In Chapter 2 of this volume, Turner et al. describe how agricultural landscape change in the Mississippi River Basin (MRB), which covers 41 percent of the contiguous United States, has caused dramatic continental-scale degradation in the form of a massive hypoxic dead zone that now appears annually in the Gulf of Mexico. Chapter 2 summarizes the large body of research that concludes that nitrogen leaching from farmland in the MRB is the leading cause of hypoxia in the Gulf, and that waters moving from the Middle Mississippi Basin of the Corn Belt states contribute the largest proportionate nitrogen loads.

Wetlands in the Corn Belt states could have mitigated downstream flooding and the export of excess nitrogen to streams, lakes, and coastal waters. More than half of U.S. wetland acres (more than 100 million acres [404,685 km$^2$]), however, have already been lost along with their ecosystem services such as flood storage and habitat values. Since colonial times, draining of land for agriculture has accounted for 80 percent of all wetland loss. With so many wetland acres gone, this trend has slowed dramatically, with agriculture accounting for 66 percent of wetlands lost after 1954. In addition, indications are that federal goals of no net wetland loss are beginning to be achieved. But the landscape legacy of past dramatic wetland losses remains—with one-third of past wetland loss in the Corn Belt states (Zedler 2003; Hansen 2006).

**Table 3-1.**    Environmental Effects of Past Corn Belt Agriculture

| *Effect* | *Degradation concerns* |
|---|---|
| **Nutrients** | Phosphorus discharge from agricultural lands is two times what it was before European settlement. The highest discharge comes from areas with a high density of concentrated animal feeding operations (CAFOs). |
| | The Mississippi River carries approximately 15 times more nitrate than any other U.S. river, and the amount has doubled since the 1950s. |
| | Current nitrogen discharge from farm fields in the Mississippi basin in certain areas is 50 times higher than presettlement rates. |
| Groundwater | The Ogallala aquifer is reported to be 50 percent depleted because of agricultural irrigation practices, and is recharging at a fraction of the rate that it is being withdrawn. |
| | More than 28 million people live within 10 miles of a polluted waterway. U.S. EPA studies have shown that 10 percent of community wells and 4 percent of rural domestic wells contained at least one pesticide. |
| | Nonpoint pollution from livestock farms has caused *Cryptosporidium* contamination in public drinking water supplies. |
| | High levels of nitrate found in drinking water have been linked to intrauterine growth retardation among women in Iowa. |
| | Studies conducted in Iowa show that more than 18 percent of private rural drinking water wells were contaminated with nitrate in levels higher than the standards set by the U.S. EPA as safe drinking water (10 ppm). |
| **Sedimentation** | Erosion and sedimentation from Corn Belt agricultural lands are causing damage in the Mississippi watershed with costs that average $2–$8 billion per year. |
| | Erosion rates are known to increase by 100 times after land is cleared of indigenous plants. |
| Habitat | Ninety-five percent of the original prairie, savanna, and woodland vegetation in the Mississippi watershed has been converted to agricultural uses. |
| | American agricultural uses have contributed to 80 percent of all wetland loss since European settlement, with one-third of that wetland loss occurring in Corn Belt states. |
| **Wildlife** | Monarch butterflies utilize the Corn Belt as a migratory stop, but there is concern about survival rates as their habitat (common milkweed) is eliminated by agricultural practices. |
| | The Corn Belt includes the global flyway for 60 percent of all North American bird species and it provides habitat for more than 300 state and federally listed endangered or threatened animal species. |
| **Climate and weather processes** | Agricultural production in the United States contributes 10 percent of GHG emissions, primarily methane, carbon dioxide, and nitrous oxide. |
| | More than 80 percent of ammonia emissions in the United States stem from agricultural production. |
| | Draining of wetlands and plowing of grasslands in the Corn Belt have led to significant increases in atmospheric carbon dioxide. Carbon content in soils has been reduced to one-third less than it was in its native condition. |

**Table 3-1.**   Environmental Effects of Past Corn Belt Agriculture (continued)

| *Effect* | *Degradation concerns* |
| --- | --- |
| **Climate and weather processes** | Wetland protection or restoration within the Mississippi River watershed could have helped to minimize the catastrophic flood of 1993 by capturing and storing storm water. |

*Sources*: Dahl (1990); Kross et al. (1993); Munger et al. (1997); Pimental et al. (1997); Zucker and Brown (1998); Guru and Horne (2000); Gollehon et al. (2001); Mitsch et al. (2001); Reilly et al. (2001); Weyer (2001); Snyder and Bruulsema (2002); The Heinz Center (2002); Heimlich (2003); TNC and NatureServe (2003a, 2003b); Turner and Rabalais (2003); Neumann et al. (2005); Shepard (2005); and Hansen (2006).

Biodiversity losses accompany water quality losses within the Corn Belt and downstream along the Mississippi and into the Gulf. While 20 percent of the larger MRB remains in native habitats, the Corn Belt's northern tall grass prairie region retains only 5 percent of the area once in native habitats (Gagnon et al. 2004). More than 70 native species have been extirpated in Iowa, and amphibians, birds, and mammals are currently threatened throughout the Corn Belt (Noss and Peters 1995). Plant pollinators such as butterflies, other insects, and birds are at risk, and aquatic habitats for fish, amphibians, birds, and insects are widely compromised by elevated levels of phosphorous and sediment in wetlands, streams, and lakes, as well as by hypoxia (Woltemade 2000; Beeton 2002).

In the Corn Belt today, the ecosystem services of indigenous ecosystems and fishable, swimmable waters have become rare and are at risk (The Heinz Center 2003). In public perceptions, these compromised environmental realities could replace traditional images of Corn Belt rural landscapes in harmony with nature, of a wholesome food system, and of an agrarian way of life (Nassauer 1997).

## Past Policies and Support for Production

Even though some federal agricultural programs have supported healthy rural communities, soil and water conservation, and protected habitats, in practice these goals have taken a backseat to production support. Especially in recent decades, policy has led U.S. farmers to expect to receive federal payments based on the amount of land in their production base for certain crops or on the amount of certain commodities produced.[1] Not surprisingly, federal payments have changed the structure of U.S. agriculture and rural communities. More surprising are the effects—subsidies have been concentrated among larger farmers and farmland owners; mid-size farms have steadily been lost (Roberts and Key 2003; Hoppe and Banker 2006); and population has declined in agricultural areas, with the associated demise of community institutions and services (Flora 2001). Furthermore, this approach has led to overproduction of subsidized commodities such as corn and soybeans, far above market demand (Babcock 1999, 2002). From the perspective of other nations, this distorts trade and puts agricultural markets elsewhere, particularly in the developing world, at a disadvantage (Babcock 2002; Hanrahan and Zinn 2005).

The subsidized crops also require high fertilizer inputs and efficient surface drainage, which speeds nutrient-rich runoff to the Mississippi River and ultimately contributes to

---

1.  These subsidies have taken many forms, such as direct subsidies, emergency payments, crop insurance, and countercyclical payments, among others. For details see CBO (2005).

the dead zone in the Gulf (Claassen et al. 2004; Chapter 2 in this volume). Ironically, agricultural policies that were intended to benefit the U.S. public may actually have undermined the environmental quality of the continent and its waters, contributing to flooding, unsafe drinking water, reduced and unsafe fisheries, and impoverished ecosystems (USDA NRCS 2006).

Although conservation has been a longtime aim of farm policy, even soil conservation, a major component of early farm program funding, has been dwarfed by production subsidies in recent decades. Figure 3-2 reports average federal conservation funding expenditures (white in the chart) relative to total farm payments (dark gray), initially by decade, and then by five-year increments, from 1930 to the present. During the dust-bowl days of the mid-1930s, conservation expenditures were a high proportion of the total funding of farm programs, averaging about 40 percent of total expenditures. Beginning in the 1960s, though, the proportion of federal investment in conservation incentives fell, reaching a low of about 4 percent of farm program funding in the 1980s. In addition, conservation programs and implementation rules have changed frequently, making it difficult for farmers to anticipate or plan how to use these programs for their economic benefit and making it risky for farmers to invest in longer term crop or stewardship practices that could have more lasting environmental benefits.

Today, although production subsidies still vastly exceed conservation spending, overall conservation spending has increased. Just under 12 percent of total farm spending has been directed at conservation in the past five years (Womach et al. 2006).

## Conservation Programs After 1985

Important policy changes to enhance environmental benefits began with the Food Security Act of 1985 (Claassen et al. 2001; Cain and Lovejoy 2004), the "1985 farm law."

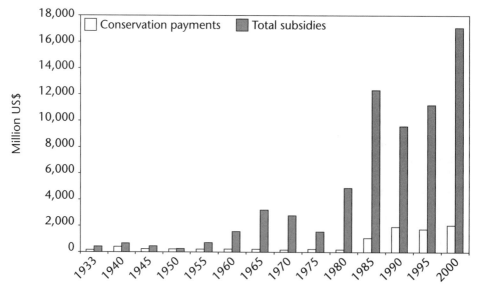

**Figure 3-2.** Conservation Payments and Total Agricultural Subsidies Paid to U.S. Farmers, 1933–2004

*Source:* USDA ERS (2006).

For example, this law introduced conservation compliance, in which farmers were eligible for certain federal payments only if they managed their wetlands and highly erodible land to achieve certain soil and water quality benefits (Claassen and Ribaudo 2006). The so-called "Sodbuster" and "Swampbuster" requirements of the 1985 law prohibited cultivation of previously unbroken sod on highly erodible land (HEL) and draining of existing wetlands. Conservation compliance also required owners of HEL to develop farm conservation plans to reduce erosion by 1995 or risk the loss of federal agricultural payments. Claassen et al. (2004) estimate that up to a 25-percent reduction in erosion can be attributed to conservation compliance.

Conservation compliance, however, has come under increasing criticism. Critics believe that enforcement has waned and that monitoring techniques may not be reliable (Claassen 2006a). Claassen (2000) reports that conservation plan violations hit a peak in 1993 when more than 850 producers lost benefits nationwide. Compare this number with the 1998 total, when only 49 producers lost benefits. Particularly relevant to water quality problems such as hypoxia of the Gulf is the success of the Swampbuster provision, which aimed to prevent drainage of wetlands for crops. Prior to the 1985 farm law, conservation programs and technical assistance actually supported draining of wetlands to increase agricultural productivity (Heimlich et al. 2000). Even though the Swampbuster compliance provision of the law is clear, its enforcement appears to have been limited, in part by its monitoring strategy (Hansen 2006). During the 12-year period from 1986 to 1998, an annual average of fewer than 100 producers were denied benefits for wetlands violations (Claassen 2000).

### The Conservation Reserve Program (CRP)

The CRP, a component of the 1985 farm law, has been more universally recognized for its success. Pairing explicit conservation aims with a goal of limiting production, the CRP retired cultivated land from active production to perennial cover (Table 3-2). Since 1985, the CRP has been the most well-funded federal conservation program, ultimately retiring more than 36 million acres (145,686 km$^2$) of formerly cultivated cropland (10 percent of total U.S. cropland), nearly all in ten-year contracts (Color Figure 28; Claassen 2006b; Hellerstein 2006). Although the conservation benefits of the CRP are widely recognized, it was initially managed to control supply by stopping production on as much HEL land as possible (Reichelderfer and Boggess 1988; Smith 1995; Smith et al. 2000; Goodwin and Smith 2003; Hellerstein 2006). By retiring 10 percent of cropland from production, the program indirectly raised commodity prices and ultimately increased farm income.

The CRP's focus on reducing erosion became more diffuse in 1990 when the Environmental Benefits Index (EBI) was introduced. The EBI broadened the conservation goals used to select land for the CRP to include wildlife habitat, water quality, erosion, enduring benefits in the form of tree planting or longer enrollment contracts, air quality, and whether the acreage was located within a Conservation Priority Area, with the cost of land counting as a seventh factor. Under the broader conservation goals of the EBI, new CRP enrollments were less concentrated in the Corn Belt than under previous criteria (Hellerstein 2006).

Many measurable environmental benefits have resulted from the CRP. Soil erosion rates have dropped significantly. Feather et al. (1999) estimate that the CRP has generated about $0.5 billion per year in the value of enhanced freshwater recreation, pheasant hunting, and wildlife viewing. These environmental benefits, though, come at a relatively high cost—the $1–$2 billion per year that funds the CRP. Some believe that the CRP has had disproportionate negative effects on rural economic activity in specific locales where the amount of land taken out of production is concentrated. But research has shown these effects to be small, not widespread, and transitory (Sullivan et al. 2004; USDA NRCS 2006).

Benefits of the CRP may be lost as ten-year enrollment contracts expire. The initial round of CRP contracts has expired, and some of that acreage was not reenrolled. For example, Sullivan et al. (2004) report that only about 12 million acres (48,562 km$^2$) of the 21 million (84,983 km$^2$) that expired in 1997 were reenrolled in that year's signup. Some landowners simply did not wish to reenroll; others found that their land did not compete favorably under the broader environmental goals of the EBI criteria. Another decisive reenrollment period is 2007, when more than 16 million acres (64,749 km$^2$) of CRP contracts are set to expire. Another 12 million (48,562 km$^2$) will expire by 2010, meaning that 28 million of the 33 million acres (113,311 of 133,546 km$^2$) enrolled in the program will have expired by 2010. Overall, because the CRP has effectively reduced erosion, improved water quality and hydrologic regimes, and created habitat, it presents a timely opportunity for present and future policymakers to build on past successes.

### Subsequent Innovations: Decoupling and Working Lands Conservation

Other conservation program innovations were introduced after the CRP. Particularly notable for their relevance to future policy innovations are (1) the decoupling of subsidy payments from commodity production (introduced in the Federal Agriculture Improvement and Reform Act [FAIRA] of 1996; the "1996 farm law") and (2) new working lands conservation programs (introduced in the 1996 farm law and the Farm Security and Rural Investment Act [FSRIA] of 2002; the "2002 farm law").

When the 1996 farm law decoupled crop subsidy payments from production, decoupling was not part of the conservation title, but it could have been used to achieve environmental benefits. For example, the European Union (EU) has now adopted decoupling with a clear conservation purpose: To receive payments, farmers must adopt certain environmentally beneficial practices that vary by region (Hanrahan and Zinn 2005). In the 1996 farm law, though, decoupling was intended to give farmers flexibility to move away from the specific crop on which their subsidy had previously been based, through a fixed contract payment. The law gradually reduced the level of fixed payments over five years. When commodity prices declined after the law was passed, Congress passed emergency payments augmenting the original fixed contract payments in an attempt to provide countercyclical support to farmers in addition to the fixed payment. The net effect was record-breaking agricultural subsidies. Under the 1996 law, the cost of federal agricultural subsidies rose from $7.3 billion in 1996 to $23.5 billion in 2000. When the 2002 law was passed, decoupled contract payments were reduced, and mechanisms from previous farm laws were added back in to provide stronger countercyclical commodity price protection for producers in addition to the decoupled income support.

Two important working lands conservation programs have used different strategies to target funds to locations where relatively high multiple environmental benefits can be achieved at low cost (Table 3-2). In 1996, the Environmental Quality Incentives Program (EQIP) was introduced in an effort to attain environmental benefits from land still in production and focus resources where the greatest environmental benefits could be achieved. EQIP consolidated previously existing cost share and technical assistance programs, most notably the Agricultural Conservation Program (Heimlich 2000), and the EQIP approach to cost sharing has varied considerably over the life of the program (Claassen 2006b). Under the 2002 farm law, most EQIP funding has been designated to address livestock-related resource concerns, including cost sharing for manure management.

The 2002 farm law aimed for more environmental benefits by introducing the Conservation Security Program (CSP; Table 3-2), which was designed to reward and encourage conservation on working lands and complement the CRP land retirement approach. Going beyond the EQIP incentive of partial federal payment for the cost of installing conservation practices, CSP compensates farmers and farmland owners for a broader range of conservation practices than ever before. The 2002 law envisioned the CSP as the first U.S. large-scale "green" payment program for agriculture. After the law was implemented, though, analysis by the Congressional Budget Office indicated

**Table 3-2.**   Recent Federal Agricultural Conservation Programs That May Be Important to Future U.S. Policy

| Program element | CRP | EQIP | CSP |
|---|---|---|---|
| **Intent** | Retired cultivated land from production and created habitat with perennial cover. Since 2002, increased integration into working agricultural land pattern | Working land cost sharing of conservation practices. Focuses funding on serious problems for air, soil, water quality, and habitat protection. Since 2002, most funding has been for livestock-related practices | Working land whole farm perspective: reward farmers for previously adopted conservation practices as well as new practices to improve air, soil, water quality, energy conservation, habitat |
| *Year implemented* | 1985 | 1997 | 2004 |
| **Acres affected** | 36 million acres (145,686 km²) of formerly cultivated cropland; 28 million acres (113,311 km²) are scheduled to expire before 2012 | 94.5 million acres (382,427 km²) | 10.2 million acres (41,277 km²) enrolled. Eligibility limited to 220 of 2,119 watersheds. Sixty more slated as eligible in 2006 |
| **Environmental benefits demonstrated** | Soil erosion rates dropped, enhanced carbon sequestration, wetlands restored, increased for open space recreation including hunting | Not yet widely measured | Not yet widely measured |
| **Costs** | $1.9 billion per year | $900 million per year | $200 million per year. Estimated $9 billion per year for full national implementation |

that nationwide implementation of the program could cost almost $9 billion by 2014 (Johnson 2004). Subsequently, watersheds where farmers were eligible to apply for the CSP were severely limited. In 2005, of the 2,119 agricultural watersheds in the nation (Claassen and Ribaudo 2006), producers in only 220 watersheds were eligible. As of 2006, producers in 280 watersheds were eligible, with another 51 watersheds slated to join in 2007. Color Figure 29 shows the location and extent of the CSP-eligible watersheds as of 2006.

## Drivers and Options for Change

For some time, both national and international policymakers have seen the need for reform in U.S. agricultural policy. In this section, we summarize three key drivers that contribute to pressures for change. First, agricultural policy is increasingly viewed as costly, considering that it appears to benefit only a small segment of society. Second, Americans are increasingly interested in quality-of-life issues related to environmental quality and ecosystem health. Finally, pressures emanating from World Trade Organization (WTO) mandates demonstrate that the existing form of agricultural subsidies is not acceptable to U.S. trade partners in the long term.

Few would disagree that federal agricultural policy has been costly, with fewer and fewer Americans directly benefiting from these costs. Farmers on the roughly 2 million remaining U.S. farms make up only about 2 percent of the population, but they depend on government payments as a substantial and stabilizing part of their incomes (Feng 2002; Hoppe and Banker 2006); net federal government transactions have averaged about 20 percent of net cash farm income since 2000 (Claassen et al. 2004; Hoppe and Banker 2006). As noted earlier, only a small fraction of that support is for conservation programs; further, a relatively small share of U.S. farms (which rose to 15 percent in 2004) receive payments specifically for conservation (USDA NRCS 2006). How the broader taxpaying public sees the benefits that they derive from agricultural subsidies is likely to influence future agricultural policy.

Even though farmers are becoming few, their land management decisions affect many, and agriculture's contribution to environmental degradation and diminishing biodiversity is becoming widely recognized. At the same time, Americans are becoming more vocal in their appreciation of the value of countryside landscapes, the ecosystem services they can provide, and the damaging effects of sprawl. State farmland protection laws, which have been adopted by every state, demonstrate growing public appreciation for open space values, scenic amenity values, and historic and cultural characteristics of agricultural landscapes in the United States (Hellerstein et al. 2002). A comprehensive review of more than 60 scholarly articles that estimate the economic value of open space to the public makes clear that the public increasingly values open space. Natural areas, though, are generally more valued than farmland and open space near large cities is more valuable than open space in more distant areas (McConnell and Walls 2005). Values ranging from $23–$830/acre ($56–$2,051/ha) have been reported, indicating that the value of preserving open space can, at times, be larger than the value of the crops produced by the land.

There is also significant evidence that improvements in water quality, habitat for recreational fishing and hunting, and preservation of biodiversity are highly valued (USDA NRCS 2006). The magnitude of the values can be gleaned from a few examples from a large and growing nonmarket valuation literature. In an analysis of almost 40 wetland valuation studies, Woodward and Wui (2001) found that while per-acre values of wetlands varied tremendously based on their purpose and use, dollar values were very high for some common uses. In particular, they predict that bird watching in the wetlands they studied is valued, on average, at more than $1,200/acre ($2,965/ha). Flood control benefits of wetlands are predicted to provide nearly $400 of services per acre ($988 of services per ha).

Studies focusing on water quality improvement in lakes and streams also find large values for reduced nutrient concentrations and sedimentation. In a study of Iowa households, Egan et al. (2005) found that people view water quality as more important than distance from the lake when they choose a lake for recreation. They also found that Iowans would be willing to spend more than $14 million annually for improved water quality at a targeted set of lakes in the state. Further evidence of the public's increasing commitment to environmental restoration are major and expensive restoration efforts such those in the Chesapeake Bay and the Great Lakes watersheds, driven in part by the environmental community's insistence that the U.S. Environmental Protection Agency (U.S. EPA) enforce the mandates of the Clean Water Act.[2]

Concerns about climate change and the need for sustainable energy supplies also may influence public views of conservation programs and farm policy. Agriculture is both a significant emitter of greenhouse gases (GHG) in the United States and a potential contributor to reductions in GHG emissions (Fletcher 2005; Pew Center on Global Climate Change 2005; Table 3-1). Increased carbon sequestration by alternative agricultural practices and landscape patterns, and adoption of biofuels are two important ways in which agriculture may prove beneficial to reducing GHG concentrations (Lewandrowski et al. 2004). Each of these possibilities is foreshadowed in the scenarios described in Part 2 of this book. These scenarios examine some of the environmental and societal implications of alternative futures that increase corn production, a consequence of increasing corn-based ethanol production, (Scenario 1) compared with some that increase perennial cover (Scenario 2), particularly native grasses as part of agricultural enterprise, a plausible consequence of cellulosic ethanol production (Scenario 3). If revamped farm programs are seen to produce more sustainable fuels and meaningfully address risks of climate change, they may be more likely to garner public support.

Perhaps of equal or more importance for the shaping of future Corn Belt landscapes is increasing scientific evidence that demonstrates how agriculture can provide ecosystem services by changing farm practices and enterprises. Examples of such ecosystem services include reduced sedimentation from contour farming, conservation tillage and buffer strips, lowered flood risk and water quality improvement from wetland restoration, and the sequestration of carbon in perennial crops and agricultural soils. Numerous studies of the adoption of "best management practices" (BMPs) provide ample scientific basis for quantifying the range of improved environmental endpoints from these and several other agricultural land uses (Gagnon et al. 2004). The alternative future scenarios described in Chapter 4 (pictured in Color Figures 4–18 and mapped in Color Figures 30–43) envision such innovative practices and enterprises and measure previously unimagined levels of societal and ecological benefits that future agricultural landscapes could deliver. They show that new policy can allow agriculture to be profitable for farmers at the same time that it builds a higher

---

2. The requirements that total daily maximum loads (TDMLs) be established and steps taken toward improving water impaired by nonpoint source pollutants has been the basis of numerous lawsuits in the past decade.

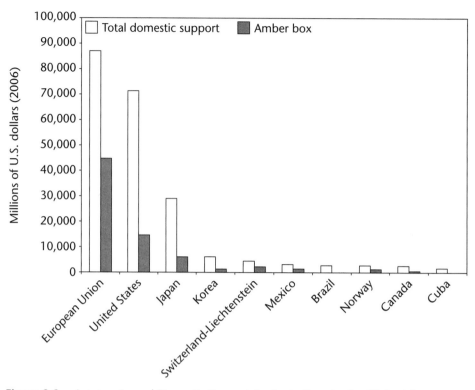

**Figure 3-3.**   Average Annual Domestic Support for Agriculture by the Highest Support Countries, 1998–Present
*Source:* CBO (2005).

quality of life and healthier ecosystems for rural Corn Belt communities as well as for ecosystems and cities extending all the way to the Gulf.

A third potential driver of change in farm policy is international pressure. Formation of the World Trade Organization (WTO) in 1994 and its predecessor, the Uruguay Round of Agreements (1992), recognized the trade-distorting effects, social equity problems, and environmental degradation consequences of national agricultural policies—not only in the United States but among the other developed nations of the world (Hart and Babcock 2002; Figure 3-3). Existing WTO rules will continue to apply regardless of the outcome of the Doha Round, which was expected to further restrict agricultural commodity subsidies.

The WTO classifies agricultural policy as "amber box," "blue box," or "green box" (Claassen et al. 2001; Hart and Babcock 2002; Moyer and Josling 2002; USDA NRCS 2006).[3] The amber box category includes all market-distorting payments, generally including payments that depend on current price and/or production levels such as commodity price supports. Under the WTO "Agreement on Agriculture," signatory countries agreed to limit such support to a reduced percentage of past expenditures. Blue box refers to direct subsidies to farmland owners or farmers based on land area in production or the amount of a commodity produced during a past, agreed-upon baseline period (85 percent of the period from 1986 to 1988 for the United States; Hart and

3.   Two additional categories, the "special and differential" box and "de minimis support," are relatively small and not central to our discussion.

Babcock 2002; Hanrahan and Zinn 2005). These subsidies are deemed less distorting than amber box, and they are not limited by WTO mandate.

Finally, green box payments are deemed to exhibit little or no market distortions. This category includes payments for environmental benefits from agricultural landscapes that the market alone may not provide, such as clean water, habitat biodiversity, or countryside landscape quality (Claassen et al. 2001; Hanrahan and Zinn 2005) as well as decoupled income support payments to farmers (Claassen 2006a), natural disaster relief, producer or resource retirement programs, and domestic food aid. Green box payments, however, are not unlimited. In the case of conservation payments, the WTO limits green box payments to the farmer's cost of undertaking the conservation effort. Although the policies in all boxes aim to provide income support for farmers, they differ in their anticipated effects on international trade (Hill 1999; Mohanty and Kaus 1999; Babcock 2001; Feng 2002).

The Organisation for Economic Co-operation and Development (OECD) nations have agreed to an agricultural policy reform agenda. This agenda includes direct support for green box public benefits—for example, a pleasing countryside or protection of ecological quality—which the OECD has determined will be more economical than current agricultural policies in achieving adequate farmer incomes as well as broader public benefits (OECD 2003). Working under a reformed Common Agricultural Policy (CAP), the EU plans to move to green box policies largely through "cross-compliance," meaning that farmers receive direct income payments regardless of the commodity produced but depending on the environmental benefits achieved. These environmental benefits range widely—from nitrate reduction to habitat enhancement to protection of landscape character (Hanrahan and Zinn 2005).

Two current U.S. conservation programs serve as strong platforms for non-trade-distorting (green box) policies: the CSP, a working lands program, and the CRP, a land retirement program (Table 3-2). In the Corn Belt, conservation payments that fall within the WTO green box could address several pressing environmental problems and deliver widely recognized benefits to both rural and urban citizens (Table 3-1). Although making changes in traditional row-cropping patterns and increasing conservation tillage alone are not likely to meet the water quality improvement goals for the Gulf described in Chapter 2 (see also Wu et al. 2004), greater water quality improvements could be achieved in the region if the menu of conservation practices and new landscape patterns is broadened. The menu could be expanded to include more of the practices and patterns discussed in the rest of this book, and the cost could be less than the total paid to the region in current farm subsidies (Kling et al. 2005). Combining both traditional and innovative BMPs in the alternative futures for Iowa watersheds (as described in Chapter 4) with an eye toward effects across the entire MRB (as described in Chapters 2 and 15) could result in a lower cost, more effective conservation policy. We detail these possibilities in the closing chapter of the book.

Agricultural conservation programs and practices need to evolve to incorporate new technology, a broader sense of ecosystem services, and attention to what the taxpaying public notices and values about rural landscapes and agriculture. The future scenarios and policy suggestions that follow in this book could help to accomplish more effective conservation program design. New programs could be designed to ameliorate local, continental, and global impacts of the past and to bring innovations to U.S. agriculture that more clearly demonstrate its benefits to society.

# PART 2

## *Assessing the Effects of Alternative Corn Belt Landscape Futures in Iowa*

*Chapter 4*

# Alternative Scenarios for Future Iowa Agricultural Landscapes

## Joan Iverson Nassauer, Robert C. Corry, and Richard M. Cruse

Changes in U.S. agricultural policy could dramatically affect the ecological quality of streams, landscapes, and communities in the Corn Belt, the Mississippi River Basin (MRB), and the Gulf of Mexico (see Chapter 1; Figure 1-1). Agricultural lands in the MRB are the primary source of the nitrogen that causes a dead zone in the Gulf of Mexico, and most of that nitrogen is coming from the Corn Belt, where the proportion of land in farms is much higher than in the MRB as a whole (USDA NASS 2004; Chapters 2 and 15 in this book). For example, 88 percent of the surface area of Iowa (31,729,490 acres; 128,404 km$^2$) is land in farms. In this chapter, we describe alternative futures for two agricultural watersheds in Iowa in order to examine how different policies, practices, and landscape patterns might produce both improved environmental quality and productive agricultural enterprises.

Farming does not have to lead to impaired water quality. Improved Corn Belt agricultural practices and landscape patterns could produce healthier drinking water; cleaner wetlands, lakes, and streams; reduced flood risks; and enhanced habitats—not only for the Mississippi but also in smaller metropolitan watersheds that lie downstream from agriculture all throughout the MRB (Chapter 3, this volume). Benefits from such improvements would accrue not only to rural ecosystems and landscapes, but also to metropolitan areas from Minneapolis to New Orleans, where flooding, water supplies, and aquatic habitats have been dramatically affected by rural watersheds upstream.

Policy changes that could enhance environmental quality may also be an effective way for the United States to meet international trade mandates that discourage commodity support payments (Chapters 3 and 16, this volume). Instead, federal policy could support agricultural practices that would produce a wide array of more tangible, broad public benefits, such as improved water quality, greater biodiversity, and more attractive rural landscapes. To examine how such public benefits might be achieved by different farm policies, in this chapter, we look at alternative futures for two second-order watersheds in Iowa—Walnut Creek in Story and Boone counties and Buck Creek in Poweshiek County. In the remainder of this part of the book, scientists report on the likely effects of these alternative futures on water quality, ecological quality, profitability, and stakeholder acceptance. In Chapter 14, Santelmann et al. offer an integrated assessment of the effects of the alternative futures in the two watersheds.

In Chapter 2, Turner et al. describe the need for alternative agricultural land uses and practices throughout the MRB, as documented by a federal, tribal, state, and local task force that conducted an integrated assessment of hypoxia of the Gulf of Mexico (U.S. EPA 2001). In Chapter 15, Scavia et al. discuss how changes in agricultural land uses and practices might be most effectively targeted to reduce hypoxia of the Gulf. They conclude that regions with the highest current nitrogen yields, including the Corn Belt watersheds we describe in this chapter, could be effective geographic targets for achieving reduced nitrogen loads to surface waters and the Gulf.

The futures we describe here put alternative policy scenarios on the ground at the scale at which agricultural practices are implemented: small, second-order agricultural watersheds. The Walnut Creek watershed extends over 13,800 acres (5,600 ha; Color Figure 30), and Buck Creek watershed covers 21,700 acres (8,790 ha; Color Figure 34). For each watershed, we describe the baseline landscape and future landscapes that could result by 2025 if each of three different policy scenarios were adopted now. Although each scenario has a different emphasis, all assume that production agriculture remains the primary land use. Each policy scenario, however, leads to a landscape in 2025 that is very different from that of 1994, the baseline year against which they are compared.

The scenarios are intended to anticipate and envision the possibility of a future that could be surprisingly different from the present. One reason to consider such surprising futures is that, in retrospect, the present is surprisingly different from the past. During the past 50 years, Corn Belt agricultural landscapes have changed in many unanticipated and not always desirable ways, especially because of their cumulative effects, such as hypoxia of the Gulf, degraded local water quality, and dramatic losses of biodiversity. While the amount of land in row crops and field sizes have dramatically increased, land uses that create more varied landscapes, such as towns, farmsteads, pastures, and woodlots, have dwindled to a degree that would have been unimaginable in 1950. For example, from 1964 to 1997, the number of farms in Iowa dropped by 41 percent to just 91,000. During the same time period, average Iowa farm size increased from 219 acres (88 ha) to 343 acres (138 ha) as well (USDA NASS 1999). The structure of farm ownership also changed. In 2002, 19 percent of Iowa farmland was owned by individuals living outside the state compared with 6 percent in 1982, and in 2002 most Iowa farmland was managed by renters (59 percent, compared with 45 percent in 1982; Duffy and Smith 2004). With fewer, larger farms, many of which are owned by people living elsewhere, small towns have been depopulated—with resulting losses in small-town businesses and civic institutions. Between 2000 and 2003, 68 of Iowa's 99 counties lost population (FDIC 2004). Compared to more traditionally scenic parts of rural America, Corn Belt communities also have attracted less recreation or tourism, and fewer new residents in retirement. Meanwhile, those who live in the Corn Belt wonder how industrial agriculture practices might be affecting their own health and quality of life (Napier and Brown 1993; Sandoz Agro Inc. 1993).

These trends could change dramatically if new policies or technologies affect the future of the Corn Belt. If policymakers are able to anticipate the consequences, such changes could be highly beneficial to communities in the Corn Belt and downstream in the Gulf. Without examining the consequences of change, policymakers may be surprised later by unintended societal and environmental costs. As a means to anticipate choices for change, this chapter poses three different alternative policy scenarios (Table 4-1) and shows landscape futures that each policy scenario could cause in the two Iowa study watersheds (Table 4-2). Although none of the scenarios is a prescription, each is intended to stimulate discussion about the consequences of different agricultural policy choices. The scenarios are prospective policy aims that could lead to the landscape futures we describe here. They are not forecasts or predictions; instead, they are plausible goals for policymaking.

Agricultural policy must respond to several constituencies and goals, some of which may seem contradictory. To explore how apparently divergent policy aims might actually achieve overlapping, multiple ecological and societal benefits, we chose three very different leading goals for the scenarios we describe in this chapter (Table 4-1):

- Scenario 1— maximizing agricultural commodity production
- Scenario 2—improving water quality and reducing downstream flooding
- Scenario 3—enhancing biodiversity within agricultural landscapes

The shaded boxes in this chapter describe the rules for mapping the landscape futures for each scenario.

## The Study Watersheds

Based on the alternative policy scenarios (Table 4-1), we developed land allocation models and used geographic information systems (GIS) to map landscape futures (Table 4-2) for the two Iowa study watersheds. We displayed these futures in the maps and digital imaging simulations (DIS), seen in the color figures of the book. We chose the study watersheds, Walnut Creek and Buck Creek, because they exemplify different soil and relief conditions (Nassauer et al. 2002).

Walnut Creek watershed (Color Figure 30) falls entirely within the thick, productive glacial till soil of the Des Moines Lobe of the recent Wisconsin glaciation (approximately 12,000 before the present [BP]). Before European settlement in the mid-19th century and construction of tile drainage systems beginning early in the 20th century, the watershed was dominated by ephemeral prairie wetlands (Color Figure 31) like much of the MRB (Chapter 2, this volume). It is flat (Color Figure 32) and, after drainage, has the excellent farmland typical of Story County. With an average corn suitability rating (CSR) of 77.6 (Iowa State University 2004; Color Figure 33), nearly all the watershed (83 percent) was planted in corn and soybeans during the baseline year, 1994.

Buck Creek watershed (Color Figure 34) was glaciated earlier, more than 30,000 years ago, and has steep slopes (Color Figure 36) on erosive, loess-derived soils. Before European settlement, oak woodlands with prairie openings covered the watershed (Color Figure 35). At the end of the 20th century, the river bottomland and the top plateau tended to be cultivated (45 percent); grazing (20 percent) and woodlands (9 percent) remained on steep slopes. Soils are typical of Poweshiek County, with an average CSR of 65.0 (Iowa State University 2004; Color Figure 37).

Because of their different landscape characteristics, the two watersheds had very different proportions of land in the leading federal land retirement conservation

**Table 4-1.**   Key Assumptions of Alternative Policy Scenarios

| | *Baseline* | *Scenario 1* | *Scenario 2* | *Scenario 3* |
|---|---|---|---|---|
| **Policy emphasis** | Support commodity production | Support commodity production | Improve water quality and reduce flooding | Enhance biodiversity |
| **Leading land retirement conservation policies** | CRP | None | None | Permanent Bioreserve Program |
| **Leading working land conservation policies or analogs** | ACP cost sharing | EQIP cost sharing | CSP, EQIP cost sharing; water quality performance standards for farms | CSP, EQIP; incentives for perennial native crops and organic growing in networks that connect reserves and streams |
| **Leading enterprises** | Cash grain; corn and soybeans; livestock in CAFOs | Cash grain; corn and soybeans; livestock in CAFOs | Mixed cash grain livestock farming; corn, beans, oats, hay rotations; and pasture. Livestock in rotational grazing | Cash grain; corn and soybeans; native perennial plants for seed or biofuels; targeted organic production; livestock in CAFOs |
| **Rural population trend** | Losing population | Losing population at current rates; small towns and community institutions at risk | Population stabilized or slowly growing with nonfarm residents; small towns and community institutions thriving | Population stabilized or slowly growing with nonfarm residents; small towns and community institutions thriving |
| **Technology** | Conventional; adequate fuel supplies; tile drainage | Conventional; adequate fuel supplies; precision farming; tile drainage | Rotational grazing; precision farming; tile drainage | Perennial strip intercropping; possible production of alternative biofuels; precision farming; tile drainage |
| **Farmer numbers** | Losing farmers; 50 percent fewer than in 1965 | Losing farmers at current rates; 50 percent fewer farmers than baseline (1994) | Losing farmers at lower rates, but 25 percent fewer than baseline (1994) | Losing farmers at current rates; 50 percent fewer farmers than baseline (1994) |

**Table 4-1.**  Key Assumptions of Alternative Policy Scenarios (continued)

| | *Baseline* | *Scenario 1* | *Scenario 2* | *Scenario 3* |
|---|---|---|---|---|
| **Landscape character** | Large farms; tidy farmsteads; mown ditches | Fewer, larger farms; tidy farmsteads; mown ditches | More farms and farm outbuildings than in Scenario 1 or 3; variety of land covers; livestock grazing | Highly varied land covers, abundant wildlife; recreational opportunities associated with trails and bioreserves; fewer farms, but many farmsteads inhabited by nonfarmers |
| **Overall public perceptions** | Safe food; growing concerns about environmental degradation; willing to pay commodity subsidies | Boring landscape; safe food; low concern about environmental effects; willing to pay commodity subsidies | Appealing landscape; attractive for recreation and tourism; willing to pay for public environmental benefits from mixed agriculture | Appealing landscape; very attractive for recreation and tourism; willing to pay for public benefits from environmentally beneficial practices |

program, the Conservation Reserve Program (CRP), in 1994 (Chapter 3, this volume). About 16 percent of the highly dissected Buck Creek watershed was enrolled in the CRP; only 1 percent of flatter Walnut Creek watershed was enrolled (Santelmann et al. 2001). The watersheds also are different in their nonfarm population growth: Walnut Creek watershed lies in the exurban growth corridor between Des Moines and Ames, Iowa; Buck Creek watershed is not in a metropolitan growth area. For purposes of this study, though, we treated future population growth in both Walnut and Buck Creek watersheds as if it were typical of rural Iowa, which will suffer population loss in the future if current trends continue (Goudy and Burke 1994).

## Alternative Policy Scenarios

In this section, we describe the alternative scenarios (Table 4-1) and resulting landscape futures (Table 4-2) in detail. Each of the three scenarios suggests a different vision of the two study watersheds in 2025, depending on agricultural policy in the early 21$^{st}$ century. In all the scenarios, policy supports profitable agricultural enterprises on private land, and farm income is assumed to remain constant across scenarios as a result of policy. Scenario 1 assumes increasing agricultural commodity production as the leading policy emphasis (Color Figure 7); Scenario 2 assumes improving water quality as the leading emphasis (Color Figure 8); and Scenario 3 assumes enhancing biodiversity as the leading emphasis (Color Figure 9). All scenarios are compared with the 1994 baseline condition (Color Figure 6).

**Table 4-2.**   Key Characteristics of Alternative Futures for Two Iowa Watersheds

| | *Present* | *Scenario 1* | *Scenario 2* | *Scenario 3* |
|---|---|---|---|---|
| **Policy emphasis** | Produce commodities | Produce commodities | Improve water quality and reduce flooding | Enhance biodiversity |
| **Field size** | Up to 320 acres (130 ha) | Up to 320 acres (130 ha); limited only by steep slopes, public roads, and maximum combine loads | Up to 320 acres (130 ha); limited by perennial cover pastures and hay fields targeted near streams, steep slopes, public roads, and maximum combine loads | Up to 320 acres (130 ha); limited by biodiversity zone, steep slopes, public roads, and maximum combine loads |
| **Agriculture inputs** | Herbicides, pesticides, and fertilizers | Herbicides, pesticides, and fertilizers, but reduced by precision farming | Herbicides, pesticides, and fertilizers, but reduced by precision farming | Herbicides, pesticides, and fertilizers, except in the biodiversity zone and where reduced by precision farming |
| **Forest area** | Dramatically reduced; a small fraction of the 19th-century area | Decreased where land is productive for cropping | Retained as wooded pasture | Retained for agroforestry or habitat in biodiversity zone |
| **Leading land retirement conservation practices** | CRP | Stream buffers of 20 ft (6 m) | Stream buffers of 50–100 ft (15–30 m); upland storm-water detention wetlands. Off-channel storage in floodplains of 200 ft (60 m) | Permanent bioreserves; stream buffers of 100–300 ft (30–90 m) |
| **Working lands conservation practices and analogous programs** | Partial adoption of conservation tillage and other BMPs | EQIP; complete adoption of no-till practices, BMPs, and precision farming | EQIP; CSP; complete adoption of no-till practices and BMPs. Network of storm-water detention filter strips and tile drainage discharge ponds in native plants. Perennial pasture cover, including maintenance of woodland pasture; livestock fenced from streams | EQIP, CSP; targeted biodiversity BMPs; perennial strip intercropping, organic agriculture. CAFO manure treatment to tertiary municipal standards |

**Table 4-2.**   Key Characteristics of Alternative Futures for Two Iowa Watersheds (continued)

|  | *Present* | *Scenario 1* | *Scenario 2* | *Scenario 3* |
|---|---|---|---|---|
| **Targeting** | Environment. Benefits Index (EBI) targets CRP | None | Perennial cover crops targeted to fields near streams | Native perennial strip intercropping and organic growing targeted to biodiversity zone connecting reserves and streams |

### Developing the Scenarios and Landscape Futures

We used an iterative interdisciplinary process to develop the scenarios (Figure 4-1). First, we established a controlled-membership e-mail "listserv" and gathered policy suggestions in a wide-reaching conversation among experts in agriculture, policy, and landscape ecology. Then, we gathered 23 experts in seven disciplines for a three-day intensive workshop in the demonstration watersheds. Working in interdisciplinary teams in the field, the experts tested and revised policy suggestions drawn from the listserv, developed relevant overriding scenario goals, and generated a menu of landscape pattern implications of those goals. After the workshop, we operationalized each of the three policy scenarios as land allocation models—sets of explicit, replicable, geographic information system (GIS)-based decision-making rules for making landscape patterns and determining landscape management assumptions (Nassauer and Corry 2004). We used these GIS-based rules to generate a landscape future for each scenario in each demonstration watershed. To generate the futures, we used ground-truthed 1994 land cover data (of 3 ft [1 m] resolution) drawn from the Midwest Agrichemical Surface/subsurface Transport and Effects Research (MASTER) project (Waide and Hatfield 1995) and soils data (of 2 acre [0.8 ha] resolution) from the Iowa Cooperative Soil Survey (1999) with the Iowa Soil Properties and Interpretations Database (Iowa State University 1996) to interpret these data for productivity and erosiveness. The high resolution of these data allowed us to design watershed futures that embodied individual field- and farm-scale decisions, such as best management practices (BMPs), rotations, and enterprises, as well as landscape and watershed scale processes and flows. The resulting GIS-derived maps of landscape futures were used to assess a wide range of societal and environmental effects for each policy scenario (Part 2, this volume). We also used the GIS-derived futures maps to locate landscape views that would change with each scenario. Then we digitally simulated the views associated with each scenario to allow stakeholders to see how the landscape could change (Color Figures 4, 5, 13–14, 21, 22, 24, 25, 26, 28, 29, 30–34).

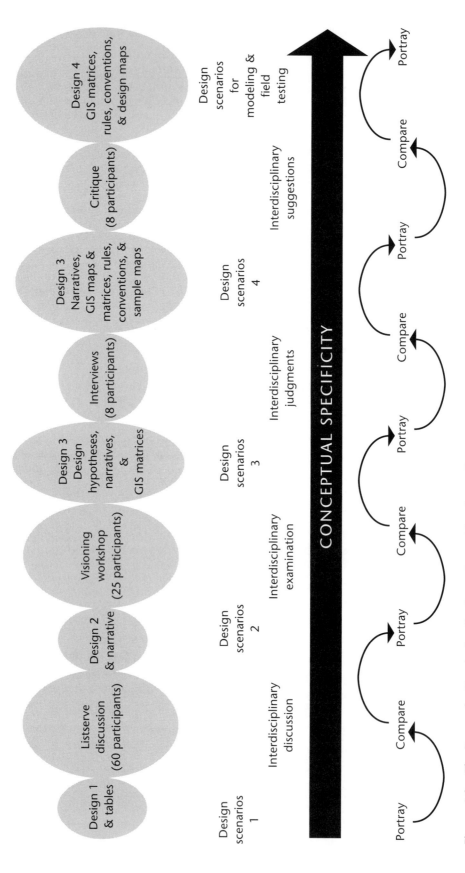

**Figure 4-1.** The Process for Developing the Scenarios and Landscape Futures

## Landscapes in 1994: A Baseline for Comparison

In 1994, the baseline year for our GIS databases (Color Figures 30 and 34), agricultural policies and programs affecting that landscape included income support to farmers determined by the following:

- past and current areas in so-called "base" commodities,
- cross-compliance requirements for farm conservation plans initiated after 1985,
- modest cost sharing for soil and water conservation BMPs under the Agricultural Conservation Program (ACP), and
- annual payments for voluntary ten-year set-aside of fields enrolled in the CRP.

Agricultural production employed intense consumption of fossil fuels and chemical and technological inputs (Cochrane and Runge 1992). Public concerns for food safety and demand for choice in food quality were increasing along with some public health concerns about water quality in agricultural landscapes, but national standards for organic food labeling had not yet been adopted (Northwest Area Foundation 1995; Lockeretz 1997). Policies to subsidize farm income were politically viable, although questions had been raised about the public benefit from farm income supports (Cochrane and Runge 1992).

## Scenario 1—Increasing Agricultural Commodity Production

The main goal in this scenario (Color Figures 38 and 39) is to increase commodity production over the short term. The primary crops are assumed to be corn and soybeans. The scenario encourages cultivation of all highly productive land and the use of conventional technologies and inputs. Consequently, in 2025, all highly productive farmland has been converted to row crops, including some areas that remained in woodland or pasture in 1994 (Color Figures 10 and 11). This scenario assumes that fossil fuel is affordable or that alternative fuels become more widely available. For example, this scenario could be driven by widespread adoption of corn-based ethanol as an alternative fuel.

Similar to policy of the past 40 years, Scenario 1 continues to favor the success of larger farms (Chapter 3, this volume). Consequently, it assumes that average farm size has increased to 640 acres (258 ha) by 2025, and that half the farmsteads occupied in the base year have been abandoned with that land cropped. No livestock enterprises are located in either of the study watersheds, although concentrated animal feeding operations (CAFOs; Color Figure 26) are clustered in a few counties of Iowa outside the watersheds (Jackson et al. 2000). Because there are no grazing livestock in Scenario 1, there are no hedgerows or fencerows that could augment the stream buffer network. No perennial crops, hay, or cover crops are planted.

With fewer, larger farms, the depopulation of the landscape has led to a loss of local schools, services, businesses, and religious institutions, although remaining farmsteads look prosperous. Lawns are tidy and large and roadsides are mown regularly and sprayed for weeds, but local biodiversity is impoverished.

The public is not attracted to visiting this homogeneous, rather empty countryside. It is widely perceived as boring "flyover" country. From a distance, however, the public perceives Scenario 1 landscapes (Color Figures 13 and 16) as environmentally acceptable, trusts the safety and quality of food produced there, and supports federal programs that subsidize and encourage large-scale industrialized agriculture.

No-till practices and precision agriculture are widely adopted under Scenario 1 because these methods conserve soil and water quality without taking even small areas of land out of production. Precision farming has encouraged more selective use of fertilizers and pesticides. Under programs much like the cost sharing available under the Environmental Quality Incentives Program (EQIP; Chapter 3), conventional BMPs of the 1990s continue in 2025, and are extended to newly cultivated land. Extensive conservation practices such as oats or hay rotations or strip cropping, however, are avoided. No-till practices may slow surface movement of sediment to streams, but the efficient tile drainage systems that enabled much of the Corn Belt to be cultivated are likely to continue to deliver nitrates, herbicides, and pesticides to local streams at a relatively high rate. Stream flows are likely to remain flashy with resulting sedimentation, flooding, and delivery of pollutants downstream and to the Gulf.

Buffer strips (20 ft [6 m]) of perennial, herbaceous, largely non-native species are required on both sides of streams, but the narrowness of the buffers and the lack of redundant connections limit their habitat potential (Henry et al. 1999). Stream buffers and dramatically diminished woodlands make up the only perennial cover remaining in this intensively farmed landscape. Native species biodiversity is limited to woodland left on the least productive soils. No reserve land program like the CRP exists. The large fields and lack of uncultivated areas create extended gaps between isolated habitats, limiting the sustainability of the already impoverished biodiversity.

The shaded box below describes the design rules we used to map the landscape futures associated with Scenario 1.

---

**Putting Scenario 1 on the Ground: Land Allocation Model Mapping Rules**

All highly productive soils are cultivated, including land that was wooded or in the CRP in 1994. This includes any area of at least 3 acres (1.2 ha) of highly productive soils as determined by local agricultural experts (CSR greater than 65 in Walnut Creek and greater than 50 in Buck Creek) that is accessible by planter and combine. Field size is limited only by steep slopes, public roads, and maximum combine loads. Combines are assumed to accommodate a pass of only 0.5 mi (0.8 km) before the combine must unload onto trucks or wagons, so sod lanes for truck or wagon access bisect 640-acre (260-ha) sections. In flat, highly productive watersheds such as Walnut Creek, the fields typically are 0.5 mi (0.8 km) wide with 320 acres (130 ha ) of corn or soybeans in a continuous rotation. Only fields larger than 30 acres (12 ha) with low soil productivity are planted in perennial crops, such as hay for sale. All fields are managed with precision agriculture. Patches of noncrop cover occur where precision agriculture identifies less productive land as small as 30 ft (9 m) wide (one combine header width) or as large as 30 acres (12 ha). These patches are mowed annually and sprayed with herbicide for weeds.

# Scenario 2—Improving Water Quality and Reducing Downstream Flooding

Shown in Color Figures 40 and 41, this scenario has related goals of improving water quality and reducing downstream flooding. Cost-share programs similar to EQIP and the watershed-based targeting of the Conservation Security Program (CSP; Chapter 3), along with clear federal performance standards for water quality from farm operations, encourage comprehensive adoption of innovative practices to improve water quality and hydrologic regimes. Many Corn Belt farms achieve farm-scale water quality outcomes by adopting integrated livestock and grain enterprises that employ perennial cover for rotational grazing. Farmers are willing to take on these management responsibilities and capital investments for meeting water quality goals because long-term federal support of operators who meet water quality performance standards is certain, and their farm incomes are secure if their farms meet environmental standards.

Livestock production with perennial cover for rotational grazing predominates on rolling land that is vulnerable to erosion.[1] Rotational grazing is managed to protect water quality (Lyons et al. 2000; Stout et al. 2000). Livestock are fenced out of streams and perennial forage crops are planted adjacent to all stream buffers (Belsky et al. 1999). Because integrated livestock operations are adopted, crop rotations that include alfalfa and hay are employed. Woodlands are maintained for grazing rather than being converted to cultivation. To reduce erosion on cropped land, farmers use conventional BMPs (e.g., no-till plowing, crop rotations, strip cropping, and filter strips).

Conventional practices are augmented by an innovative network of practices to detain storm water and sediment. On steep upland slopes (more than a 10 percent grade) in cultivated fields, a 10-ft (3-m) filter strip of native herbaceous cover is located halfway up the slope. These strips link small detention ponds to create a detention and filter "necklace" that runs parallel to the slope (Richardson and Gatti 1999). The strips are mown only annually in late summer to provide for nesting habitat. Discharge ponds are located downstream to remove nutrients and sediment from the tile and ditch drainage system. Off-channel ponds for nutrient uptake by plants are located every 1.5 mi (2.4 km) at county drain tile outlets to the stream (Woltemade 2000). Sediment and runoff from roads and roadsides is captured in off-channel ponds at all road crossings of streams. Finally, along low-gradient reaches of the stream itself, large areas (up to 200 ft [60 m] from the stream) are designed for off-channel storage. In these selected locations, constructed meanders or oxbows are planted with native species to create habitat. Continuous cover created by rotational grazing, hay and pasture (along with the comprehensive network of innovative detention and discharge ponds; Richardson and Gatti 1999; Manale 2000), traditional BMPs, and broader stream buffers (50 or 100 ft [15 or 30 m] from the center of each stream) are intended to substantially improve water quality and reduce stream flashiness and flooding.

Although Scenario 2 takes no land out of production as a conservation area or bioreserve, its practices create landscape networks of habitats and redundant connections among them (Henry et al. 1999; Dale et al. 2000; Hinkle 2000; Color Figures 10 and 12). To enhance habitat, planting within the buffer strip is modeled after the successful demonstration on Bear Creek, Iowa (Isenhart et al. 1995; Color Figure 22). Riparian woodlands are occasionally broken with herbaceous gaps less than 10 ft (3 m) long to

---

1.  As measured by Land Capability Class (LCC). As a measure of soil productivity, LCC does not assume the management inputs assumed by CSR. Consequently, LCC may reflect management inputs appropriate to rotational grazing enterprises with more validity.

create more diverse aquatic habitat than an entirely shaded stream (Henry et al. 1999). Mesic woodlands are maintained for grazing. The many patches of continuous cover in hay and pasture, innovative BMPs, selectively mown roadsides, and broader stream buffers would create more habitat for birds and small mammals—even if invasive species were not controlled.

Under Scenario 2, more farmers occupy the Corn Belt in 2025 compared with the other future scenarios because farmers are needed to manage livestock in rotational grazing. Still, there are 25 percent fewer farmers in 2025 than in 1994. But because Scenario 2 has created appealing landscapes that attract tourists, hunters, telecommuters, retirees, and second-home owners, local population has increased, and the variety and availability of local services, schools, and churches that existed in the 1990s has broadened in 2025 (Color Figures 14 and 17). Visitors and new community members contribute directly to farm incomes as well as to the economies of small nearby towns. Mindful of the local benefits of an agricultural landscape that appeals to visitors, farmers tend to manage their land for multifunctional values. Roadsides are mown only after the bird nesting and brooding season.

---

**Putting Scenario 2 on the Ground: Land Allocation Model Mapping Rules**

Rotational grazing occurs on less productive land and land near streams; corn rotations take place on highly productive land. Any field that has at least 75 percent highly productive cropland (LCC 1, 2, or 3) is assumed to be cultivated in 2025. Areas larger than 40 acres (16 ha) that are not highly productive by this standard are managed as separate parcels for rotational grazing or hay. In flatter areas with highly productive soils, fields are as large as 320 acres (130 ha). These fields are planted in a corn–soybean–oats–alfalfa six-year rotation (Waide and Hatfield 1995) and managed with no-till practices, conventional BMPs, and precision agriculture. As in Scenario 1, any area of at least 3 acres (1.2 ha) of highly productive land accessible by combine is cultivated. Woodlands adjacent to or in pasture or forage crops, however, are maintained for grazing. On all soil types, forage crops for hay or rotational grazing are planted in blocks of at least 10 acres (4 ha) adjacent to all perennial stream buffers.

---

## Scenario 3—Enhancing Biodiversity within Agricultural Landscapes

The goal of this scenario (Color Figures 42 and 43) is to enhance biodiversity in the context of agricultural production. As a means of achieving that goal, perennial grasses are grown for market enterprise. Native perennials are integral to a new system of Corn Belt bioreserves as well. This new reserve program purchases less productive land from willing sellers to create a network of habitats within an agricultural landscape. The habitat network includes perennial crops and BMPs that further enhance biodiversity. Federal policy establishes a means for public ownership and management of these permanent bioreserves, which are selected to represent indigenous ecosystem types (e.g., wetland and riparian woodland). On agricultural land, under programs similar to the existing EQIP and CSP, networks of wide stream buffers and biodiversity BMPs connect to the bioreserves. This connectivity makes native plant and animal populations more viable by allowing these species to move through agricultural lands. Reserve sites are selected to maximize heterogeneity within a broad ecosystem type (e.g., wetlands should include perennial and ephemeral conditions) and to maximize interior conditions, including core habitat of at least 640 acres (260 ha) without roads or trails. Such reserve areas are large enough to create opportunities for improving hydrologic regimes as well (Color Figures 4, 5, 15, and 18).

To manage flows across reserve boundaries, filters are located at reserve edges. Small ponds detain road and field runoff before it enters reserves. To help drivers see animals crossing roads, woodland reserves are buffered from state highways or freeways with a lower herbaceous edge that is 164 ft (50 m) wide. Recreational trails are located at the edges of reserves to allow people to enjoy them while minimizing fragmentation.

A stream corridor of continuous perennial vegetation is similar to Scenario 2, but it is wider to create a diversity of riparian and aquatic ecosystems. It extends 100 ft (30 m) from the center of ephemeral streams, 200 ft (60 m) from the center of perennial streams, and 300 ft (90 m) from stream reaches where a recreational trail runs along the riparian corridor edge. As in Scenario 2, this wide buffer captures sediment and nutrients, promotes bank stabilization, and detains storm water.

Cost sharing for innovative, biodiversity BMPs is targeted at a connecting biodiversity zone that extends 0.25 mi (0.4 km) in all directions from the boundaries of reserves and stream corridors (Color Figure 22). Within this zone perennial strip intercropping, organic crops, and agroforestry are encouraged on appropriate soils. Perennial strip intercropping (Color Figure 19) adapts strip intercropping to retain its production advantages for row crops (Cruse 1990; Exner et al. 1999) but also includes a marketable, perennial native herbaceous strip between rotating strips of corn and soybeans. It consists of 10–20 ft (3–6 m) of corn adjacent to and rotated with 10–20 ft (3–6 m) of soybeans adjacent to 40 ft (12 m) of perennials, which do not rotate. Strip widths vary within the rotational sequence to avoid continuous monoculture cropping. Annual crops within perennial strip intercropping are cultivated to control weeds or genetically modified to allow for chemical weed control. Perennial strip intercropping promotes the flow of native plant species and some wildlife along and across perennial strips. Native plant seeds or perennial biofuels, such as switchgrass, could be harvested from the perennial strip. Perennial strip intercropping, which is used in fields that have less than half their area in the most productive soils, is likely to enhance water quality by taking up nutrients, holding sediment, and detaining surface water (Chow et al. 1999; Eghball et al. 2000). More pervasive planting of perennials as market crops, or planting roadsides and gardens with native plants would further enhance water quality.

Within the biodiversity zone, organic crops, which are likely to support invertebrate biodiversity, are planted on the most productive soils. Agroforestry, including timber production and mast crops, is implemented in patches of at least 10 acres (4 ha) along streams on less productive soils. Beyond the biodiversity zone, farmers use conventional BMPs, no-till practices, and precision agriculture for row crops, much as in Scenario 1. In Scenario 3, however, small patches of less productive soil identified by precision agriculture are planted with perennial native herbaceous vegetation, extending habitat. Even beyond the biodiversity zone, strip intercropping (without a perennial strip) is now employed wherever at least half a field does not have highly productive soil. Like Scenario 1, Scenario 3 assumes that livestock are raised in CAFOs in a few locations in the state outside the study watersheds. Under Scenario 3, though, CAFO waste treatment facilities meet effluent and risk standards comparable to tertiary treatment of municipal waste. Because livestock are part of very few farm enterprises under Scenario 3, no pasture and little hay are on the land.

Farm size in Scenario 3 averages 640 acres (258 ha), the same as in Scenario 1, with the result that the number of farms decreases by about 50 percent from the baseline (1994). Compared with Scenario 1, however, Scenario 3 results in a very different, varied landscape pattern with abundant wildlife and recreational trails (Color Figure 18). These amenities and the public's perception of Corn Belt agricultural landscapes as healthy natural places attract nonfarmers as visitors and new residents, as in Scenario 2. Nearly all the farmsteads present in 1994 are inhabited in 2025, but many of these inhabitants are new residents who do not farm. Because of the new residents and additional nonfarm income sources, diverse rural services, institutions, and businesses are widely available. Farmsteads have changed to include gardens of native grasses and flowers. Roadsides, too, are planted with native herbaceous plants, and are mown only once per year in late summer to provide habitat and also prevent encroachment by trees.

---

**Putting Scenario 3 on the Ground: Land Allocation Model Mapping Rules**

Bioreserve sites are selected on soils that evolved as part of the appropriate indigenous ecosystem type (e.g., a wetland reserve on hydric soils), and selections based on soils are validated against presettlement vegetation. Reserve sites that best match the goal of 640 acres (260 ha) without roads or houses are preferred. To maximize interior conditions, adjacent sites that meet the primary selection criteria for two different ecosystems are preferred. To make the best use of local cropland, the core reserve is extended wherever adjacent land of at least 40 acres (16 ha) has soils that are relatively unproductive. Because the scale of many ecological functions exceeds the size of the study watersheds, reserves are sited near watershed boundaries to enhance cross-watershed connectivity.

To connect habitats and protect water quality, biodiversity BMPs such as perennial strip intercropping and organic growing are targeted to a biodiversity zone extending 0.25 mi (0.4 km) in all directions from the boundaries of reserves and stream corridors. Beyond the biodiversity zone, cultivated fields are extended into any area of at least 3 acres (1.2 ha) of productive soils (LCC 1, 2, or 3) that is accessible by combine. New fields are cultivated in areas of at least 40 acres (16 ha) of productive soils. Field size is limited only by steep slopes, public roads, and maximum combine loads, as in Scenario 1. If at least half the land in a cultivated field is not highly productive land (LCC 1 or 2), the field is cultivated using strip intercropping without a perennial strip.

---

## Conclusions: Future Scenarios and Future Benefits

The alternative scenarios lead to very different future agricultural landscapes that promise widely varied environmental and societal benefits. These benefits will matter to the public, to rural communities, and to farmers (Chapters 3 and 6, this volume). To assist policymakers in making choices for future policy, these benefits are modeled, measured, and compared in the remainder of this part of the book.

All of the scenarios and related futures are plausible. Each scenario builds on existing successes in U.S. agricultural policy and technology, and each alternative future draws on expert knowledge from many disciplines. None of the scenarios dictates a single approach to agricultural policy. Different combinations of the policies, land uses, and practices employed in each scenario could be implemented to achieve a future landscape that balances policy goals in a different way. Practices and technologies that we did not imagine undoubtedly are important as well. The alternative scenarios powerfully illustrate that dramatic change to affect environmental and societal benefits is possible, and that policy that conforms to broader world trade imperatives and the demands of evolving public perceptions could direct that change.

Policymakers face a number of questions. Is it possible to have a healthy U.S. agricultural economy, a healthy food supply, healthy rural communities, healthy agricultural ecosystems, healthy streams, and a healthy Gulf of Mexico? Can this be achieved while the United States helps to feed the world, aims to achieve greater energy independence, and trades equitably with other nations? Can agricultural landscapes be reclaimed as desirable places to live and delightful places to visit? Is each of these a legitimate goal of federal agricultural policy? Should we have to compromise any one of these environmental or societal goods to achieve another? The scenarios described here and the assessments in the following chapters that test their results can help policymakers answer these questions.

# Chapter 5

# Economic Implications

## Colette U. Coiner, JunJie Wu, Stephen Polasky, and Mary V. Santelmann

Agriculture dominates many Corn Belt watersheds. To be economically sustainable, an agricultural system must produce adequate income for farmers. To be environmentally sustainable, it must also provide adequate soil quality, water quality, and biodiversity. In this chapter, we investigate potential economic and environmental consequences of three alternative future landscape scenarios—one emphasizing production, one emphasizing water quality, and one emphasizing enhanced biodiversity—for two agricultural watersheds in Iowa. The three alternative future scenarios and the baseline landscapes for Walnut Creek and Buck Creek watersheds are described in detail in Chapter 4 and illustrated in Color Figures 38–43.

### Study Area and Alternative Landscape Futures

In Walnut Creek watershed in Boone and Story counties, Iowa, rainfall averages 32 in. (821 mm) per year, and soils range from silt loams to clay loams. The watershed topography is flat, with slopes ranging from zero to nine percent. Commercial agriculture dominates the watershed, with more than 83 percent of current land cover in crops, primarily corn and soybeans (Table 5-1a). Buck Creek watershed in Poweshiek County has slightly higher rainfall (36 in. [912 mm]) and rolling topography. Land cover in the Buck Creek watershed is more diverse than that of Walnut Creek, with only 45 percent of the area in corn and soybean row crops (Table 5-1b).

Several types of farming methods, ranging from conventional tillage to conservation tillage to strip intercropping, are currently used in the study watersheds. In our baseline analysis of the current landscape, we assumed that conventional tillage is the dominant method. Three cropping systems characterize the study watersheds. In both watersheds, two-year rotations of corn and soybeans are common. In rolling Buck Creek watershed, other typical cropping systems include oats and alfalfa in a four-year rotation, in which oats are planted as a companion crop and harvested for grain in the first year and alfalfa is harvested three times per year in Years 2, 3, and 4. Grass hay on a four-year rotation is harvested once a year. Coiner et al (2001) describes these cropping systems in more detail.

In all three future scenarios, no-till conservation tillage and current best management practices (BMPs) are assumed to be widely adopted. In Scenario 1, the primary

**Table 5-1.** Land Use under Alternative Futures

### a. Walnut Creek Watershed

| Production enterprise | Baseline | | Scenario 1 | | Scenario 2 | | Scenario 3 | |
|---|---|---|---|---|---|---|---|---|
| | Acres (hectares) | % | Acres (hectares) | % | Acres (hectares) | % | Acres (hectares) | % |
| Corn–soybean | 10,314 (4,174) | 81 | 11,122 (4,501) | 88 | 0 | 0 | 4,248 (1,719) | 33 |
| Grain oat–alfalfa | 222 (90) | 2 | 0 | 0 | 0 | 0 | 0 | 0 |
| Grass hay | 109 (44) | 1 | 2 (1) | <1 | 890 (360) | 16 | 5 (2) | 0 |
| C–S–C–S–O–A Rotation | 0 | 0 | 0 | 0 | 8,095 (3,276) | 64 | 0 | 0 |
| Alfalfa | 0 | 0 | 0 | 0 | 988 (400) | 8 | 0 | 0 |
| Strip intercropping | 0 | 0 | 0 | 0 | 0 | 0 | 5,120 (2072) | 40 |
| Perennial cover non-crop | 0 | 0 | 316 (128) | 2 | 746 (302) | 6 | 0 | 0 |
| Pasture | 294 (119) | 2 | 0 | 0 | 2,044 (827) | 0 | 0 | 0 |
| All other land (woodland, fencerows, riparian, wetlands, etc.) | 1,757 (711) | 14 | 1,255 (508) | 10 | 904 (366) | 12 | 3,328 (1,347) | 27 |
| Total land | 12,696 (5,138) | 100 | 12,696 (5,138) | 100 | 12,696 (5,138) | 100 | 12,696 (5,138) | 100 |

### b. Buck Creek Watershed

| Production enterprise | Baseline | | Scenario 1 | | Scenario 2 | | Scenario 3 | |
|---|---|---|---|---|---|---|---|---|
| | Acres (hectares) | % | Acres (hectares) | % | Acres (hectares) | % | Acres (hectares) | % |
| Corn–soybean | 9,447 (3,823) | 43 | 13,039 (5,277) | 60 | 0 | 0 | 217 (88) | 1 |
| Grain oat–alfalfa | 0 | 0 | 0 | 0 | 0 | 0 | 0 | 0 |
| Grass hay | 0 | 0 | 0 | 0 | 0 | 0 | 0 | 0 |
| C–S–C–S–O–A Rotation | 0 | 0 | 0 | 0 | 2,627 (1,063) | 12 | 0 | 0 |
| Alfalfa | 875 (354) | 4 | 1,458 (590) | 7 | 8,997 (3,641) | 41 | 0 | 0 |
| Strip intercropping | 0 | 0 | 0 | 0 | 0 | 0 | 13,423 (5,432) | 62 |
| Perennial cover non-crop | 3,440 (1,392) | 16 | 4,742 (1919) | 22 | 919 (372) | 4 | 0 | 0 |
| Pasture | 3,010 (1,218) | 14 | 0 | 0 | 4,174 (1,689) | 19 | 0 | 0 |
| All other land (woodland, fencerows, riparian, wetlands, etc.) | 5,024 (2,033) | 23 | 2,555 (1,034) | 12 | 5,073 (2,053) | 23 | 8,154 (3,300) | 37 |
| Total land | 21,794 (8,820) | 100 | 21,794 (8,820) | 100 | 21,794 (8,820) | 100 | 21,794 (8,820) | 100 |

objective is to increase agricultural production, and corn and soybean production is expanded at the expense of other land covers as shown in Table 5-1 (a, b). Relative to the baseline, cultivated land is increased. In Scenario 2, water quality is the dominant goal, and grazing livestock are introduced. Consequently, cropland is planted with a corn, soybean, oats, and alfalfa rotation of six years that incorporates conservation tillage (Coiner et al. 2001). Rotational grazing occurs on land with lower corn suitability ratings (CSR) that is fenced from streams. Buffer strips of at least 50 ft (15 m) are established along stream corridors (Color Figure 17). Color Figures 8, 12, 14, and 17 show land cover patterns for this scenario. In Scenario 3, the main objective is to enhance biodiversity, and a large bioreserve is established in each watershed (Color Figures 5 and 15). In addition, innovative farming practices (e.g., perennial strip intercropping) are targeted to connect bioreserves and riparian corridors (Color Figures 9 and 18). Scenario 3 utilizes three additional cropping systems. The first system consists of corn and soybeans strip intercropped with a grain oat crop (Color Figure 19). The second strip intercropping system includes the use of native prairie grass rather than oats (Color Figures 9 and 18). The third system includes certified organic farming, in which no commercial fertilizer and chemicals are used within the biodiversity target zone.

## Comparing the Alternative Futures to a Baseline

We examined economic and environmental implications of the alternative futures for each scenario compared to a baseline of land use in 1994. The environmental effects analyzed were nitrate-nitrogen (nitrate-N) runoff and leaching, and wind and water erosion. For comparison, we also present data from Vaché et al. (2002) on nitrate and sediment export from the Iowa study watersheds for the same alternative futures. The economic impact analyzed was total return to land (RTL), calculated as the total revenue minus total cost (except land cost). To estimate RTL under various alternative land use/crop management practices, we first estimated yields for various crops under the future alternatives. We used crop enterprise budgets from the Iowa State University Extension Service to calculate production costs.[1] County-level prices from 1987 to 1997, which were collected by the U.S. Department of Agriculture (USDA) National Agricultural Statistics Service (NASS), were used to calculate total revenues, indexed to 1998 dollars using the indexes of prices received by farmers for all farm products. Little information was available on prices or yields of native prairie grass production for seed or on possible prices for corn or perennial grasses as biofuels when we conducted this analysis. Consequently, we estimated returns for native prairie grass seed production based on yield and price information from a private producer in central Iowa, and we did not estimate returns for corn or perennial grasses as biofuels.

Both yield and some environmental effects for the Walnut Creek watershed were derived using the Interactive Environmental Policy Integrated Climate (i_EPIC) model, formerly the Erosion Productivity Impact Calculator (EPIC; Williams et al. 1988; Sharpley and Williams 1990). The EPIC model is a field-level simulation model developed to estimate management impacts on agricultural production and soil and water resources (Williams et al. 1988). Its major modeling components include weather simulation; plant growth; nutrient cycling; hydrology; erosion and sedimentation; pesticide fate;

---

1. Paul Mitchell of Iowa State University supplied the dates and types of tillage operations that occurred for each crop enterprise. Mike Duffy, Alan Vontalge, and John Lawrence from the Iowa State University Extension Service supplied 1998 Iowa state crop enterprise budgets.

soil temperature; and management of the plant environment by means of tillage, fertilization, irrigation, and conservation practices.

In our study, we used the i_EPIC model, which was developed by Todd Campbell of Iowa State University (Campbell 2000). This model uses information on weather, soils, crops, and production practices to simulate crop yields and a number of environmental indicators including nitrate-N leaching, nitrate-N runoff, wind erosion, and water erosion. As in previous studies that used EPIC, we calibrated i_EPIC to better reflect crop yields for soil types in the study watersheds (Coiner et al. 2001).

For each of the three alternative scenarios and the baseline landscape, a geographic information systems (GIS) database was created (as described in Chapter 4) that defined land cover and land management practices for each parcel in the watershed. Soil and crop yield information for the watershed came from the Iowa Soil Properties and Interpretation Database (ISPAID).[2] This database provided soil types, the number of layers with specific information on each layer (e.g., horizon depth, bulk density, particle percentages, pH, and organic matter content), and the depth of the water table. Todd Campbell supplied additional information on soils occurring in the Walnut Creek watershed. The soils information was combined with the land cover data for each alternative future to create a combined land cover–soils GIS database for the watershed.

Each scenario defined crop and basic production practices. Detailed production practice information such as the types and dates of tillage operation and the amounts of pesticide and fertilizer applications for each crop were derived from the 1998 Iowa state crop enterprise budgets developed by Iowa State University Extension Service and other sources. We conducted a 30-year i_EPIC simulation run for each field in the Walnut Creek watershed. The 30-year simulation runs allowed us to capture the effect of crop management practices on soil quality, including the percentage of organics in the soil. The resulting changes in soil quality are reflected in changes in simulated crop yields. Each run provides daily estimates of nitrate-N leaching, nitrate-N runoff, wind erosion, and water erosion. Nitrate-N leaching is measured as the quantity of nitrate-N leaving the root zone, and nitrate-N runoff is measured as the quantity of nitrate-N leaving the field via surface runoff. We used the 30-year averages of simulated nitrate-N runoff, nitrate-N leaching, water erosion, and wind erosion in the analysis.

EPIC has been validated and calibrated for a wide variety of conditions, particularly for those prevalent in this study region (Jones et al. 1985; Williams et al. 1988). EPIC has also been shown to be a reasonable predictor of nitrate-N runoff and leaching losses on several sites (Jones et al. 1985; Mapp et al. 1994; Chung et al. 1999). These results suggest that EPIC can provide a relatively accurate estimate of the long-term average of water and nitrate-N leaching losses for the cropping systems evaluated in this study. We made no attempt, however, to model the soil and nitrate-N movements after they leave the field in surface runoff or leach below the root zone. In essence, our estimated nitrate-N loadings represent a "worst-case" scenario because transport processes tend to reduce the amount of nitrate-N that actually reaches surface or groundwater sources As described in Chapter 7, Vaché et al. used SWAT (Soil and Water Assessment

---

2.   ISPAID was created by the Iowa Cooperative Soil Survey (ICSS). The ICSS is a partnership between the Iowa Cooperative Extension Service and the Iowa Department of Agriculture and Land Stewardship, Division of Soil Conservation, and the U.S. Department of Agriculture (USDA) Natural Resources Conservation Service. The general purpose of the ICSS is to coordinate the collection, compilation, interpretation, publication, dissemination, and use of soil surveys in Iowa.

Tool) to model transport processes, including nutrient uptake and transformation, for water quality in these same alternative futures.

For the more rolling landscape of Buck Creek, we used i_EPIC only to estimate yields for calculating financial profitability across scenarios. The diverse agricultural enterprises designed for the alternative futures in Buck Creek (such as livestock operations and agroforestry) precluded the use of i_EPIC for modeling some environmental effects at the watershed scale. In the alternative futures for the Buck Creek watershed, the only alternative future in which more than half of the watershed area was planted to corn and soybean row crops (the enterprises for which our i_EPIC calibrations were satisfactory) was that resulting from Scenario 1. In the baseline and alternative future landscapes of Scenarios 2 and 3, area in row crops ranged from 45 to 12 and 1 percent, respectively (although in Scenario 3, 62 percent of the Buck Creek watershed area was planted to a strip intercropping system that included corn and soybeans as well as permanent strips of native perennials). In Scenario 2, extensive livestock operations were expected to make up a substantial proportion of the agricultural enterprises in the watershed, and in Scenario 3, alternative cropping, agroforestry, and specialized crops were expected to occur in large areas of the watershed. The nature and spatial extent of these combinations of enterprises precluded watershed-level modeling with i_EPIC for estimation of nitrate-N runoff and leaching and soil and wind erosion because the necessary data for calibration within the model were lacking. Also, in contrast to models such as SWAT, i_EPIC does not route flow through landscape elements designed to remove sediment and nutrients, such as created wetlands and sediment detention ponds, which were particularly important components of the landscape futures for Scenario 2 and 3 in the rolling Buck Creek watershed. In Table 5-2, then, we present the results of the financial profitability and environmental effects based on results from the i_EPIC model, as well as water quality modeling results from Vaché et al. (2002), who used the SWAT model (Arnold et al. 1995) to compare water quality impacts from these same alternative futures and the baseline landscape for nitrate-N, sediment export, and annual stream discharge.

## Results Suggest Changes May Improve Performance

Some changes in management practices in the future scenarios improved both economic and environmental performance. For example, adopting conservation tillage (a component of all future scenarios) increased RTL and reduced soil erosion. In others, there were trade-offs between improving different economic or environmental measures. In addition, some changes had little economic impact, but could result in trade-offs between different environmental objectives such as erosion and nitrate leaching.

### *Agricultural Profits*

The first column of Table 5-2 shows the total RTL for the baseline landscape and the three alternative future scenarios. The total RTL under the baseline landscape is about $1.5 million for the Walnut Creek watershed and $1.4 million for the Buck Creek watershed. The corn–soybean rotation was the most profitable of the three cropping systems, followed by grain oats–alfalfa (Coiner et al. 2001).

Under Scenario 1, almost all cropland was assumed to be in corn and soybean production, and some areas of noncropland were converted to cropland. Conservation

**Table 5-2.** Profitability and Environmental Impacts of the Practices Associated with the Baseline Landscape and Alternative Futures[a]

| Landscape | Return to land, $/yr | N runoff, lb/yr (kg/yr)[c] | N leaching, lb/yr (kg/yr) | Nitrate export,[b] lb/yr (kg/yr) | Water erosion, st/yr (t/yr) | Wind erosion, st/yr (t/yr) | Sediment export,[b] st/yr (t/yr) |
|---|---|---|---|---|---|---|---|
| | | | *a. Walnut Creek* | | | | |
| Baseline landscape | 1,549,665 | 52,437 (23,835) | 97,088 (44,131) | 139,509 (63,413) | 34,711 (35,269) | 34,108 (34,656) | 524 (532) |
| Scenario 1 | 1,730,988 | 39,428 (17,922) | 121,777 (55,353) | 150,286 (68,312) | 10,319 (10,485) | 10,396 (10,563) | 448 (455) |
| Scenario 2 | 1,176,091 | 16,432 (7,469) | 86,962 (39,528) | 37,013 (16,824) | 7,355 (7,473) | 4,549 (4,622) | 174 (177) |
| Scenario 3 | 1,522,612 | 22,757 (10,344) | 131,314 (59,688) | 59,442 (27,019) | 5,262 (5,347) | 6,822 (6,932) | 234 (238) |
| | | | *b. Buck Creek* | | | | |
| Baseline landscape | 1,353,550 | n/d[c] | n/d | 119,592 (54,360) | n/d | n/d | 1,647 (1,673) |
| Scenario 1 | 1,940,854 | n/d | n/d | 142,426 (64,739) | n/d | n/d | 1,482 (1,506) |
| Scenario 2 | 846,938 | n/d | n/d | 43,091 (19,587) | n/d | n/d | 907 (922) |
| Scenario 3 | 2,120,568 | n/d | n/d | 55,636 (25,289) | n/d | n/d | 1,039 (1,056) |

[a] Data reported here were modeled using i_EPIC, except for nitrate export and sediment export. The nitrate export values calculated by Vaché et al. (Chapter 7, this volume) are comparable to the sum of the nitrate runoff and nitrate leaching values calculated by i_EPIC because Vaché et al. assumed that most of the nitrate leaching into the subsurface would be captured and removed in tile drainage flow. Water and wind erosion values calculated by i_EPIC represent the sum of erosion from all fields in the watershed, and differ from annual sediment export (sediment transported from the watershed by the stream) calculated by SWAT because not all of the soil that moves from a field leaves the watershed.

[b] SWAT data from Vaché et al 2002.

[c] No data were available for calibration of practices that characterized Buck Creek watershed under these scenarios.

tillage, which saves fuel, labor, and machinery costs, was employed on all row crops, and areas of least profitable soil types were not cropped because precision farming is used to identify soils with low CSRs. With the increase in total crop area and adoption of no-till and precision-farming practices, total RTL increased for both watersheds (Table 5-2). The RTL for both corn–soybean and grass hay production was higher in Scenario 1 than in the baseline landscape.

In Scenario 2, total agricultural land was increased compared to the baseline landscape because of a large increase in grass hay area. The total RTL, however, was reduced compared with the baseline (Table 5-2). In this scenario, the corn–soybean rotations are extended to six years. In the final two years of the rotation, companion crops of oats for "haylage" and alfalfa are planted. For most fields, the oat–alfalfa phase yields low return to land. After only one full growing season, the alfalfa is not producing at its full potential.

Agricultural practices envisioned under Scenario 3 generate total RTLs of $1.5 million and $2.1 million for Walnut Creek and Buck Creek watersheds, respectively. This total is only slightly below the baseline landscape for Walnut Creek (Table 5-2a) despite the fact that more than 988 acres (400 ha) have been taken out of production agriculture. This scenario generates the highest average RTL among the three scenarios and the baseline ($162/acre [$400/ha]) in the rich soils of Walnut Creek watershed) because of the high estimated returns in the corn–soybean–native prairie grass strip intercropping. This high RTL, however, depends on the assumption that native grass seeds or another enterprise market like cellulosic ethanol will continue to command a relatively high price for perennials. If the price of native grass seeds were assumed to be half of the current level, the return to land for the corn–soybean–native prairie grass strip intercropping enterprise would be reduced by approximately 35 percent. On the other hand, this model did not consider a biofuel market for perennial grass, and that might have enhanced the profitability of perennials in Scenario 3.

In Scenario 3, the organic enterprise uses the same six-year crop rotation (C–S–C–S–O–A) as in Scenario 2. The RTL for this rotation is higher in this scenario than in Scenario 2, however, because in organic farming no artificial chemicals and fertilizer were applied, reducing production costs. Because of the lack of long-term price data for organically grown crops, we used prices for conventionally grown crops instead. This tends to underestimate the returns for organics, which currently enjoy a price premium.[3] As more producers switch to organic production, though, prices for organic crops may approach those for conventionally grown crops.

In Table 5-2(b), we compare the RTL under the baseline landscape with the three future scenarios for the Buck Creek watershed. Buck Creek has more hilly terrain and is less fertile on average than Walnut Creek. As in Walnut Creek, Scenario 1, which emphasizes production agriculture, generates an increase in returns compared to the baseline. Because of its hilly terrain, Buck Creek's baseline landscape includes a significant fraction of land in pasture or in the Conservation Reserve Program (CRP). A large part of this land is converted into corn and soybeans in Scenario 1. Because this land is less

---

3.  A 50 percent increase in organic crop prices would result in an 86 percent increase in the average return to land for this enterprise (from $145 to $269/acre [$358 to $665/ha]).

fertile than land currently devoted to corn and soybeans, the average RTL per hectare for a corn–soybean rotation was lower than under the baseline, although total returns increased by nearly $600,000 because of the large increase in land area in crops. Under Scenario 2, less profitable management practices and cropping patterns resulted in a large decrease in returns. In Buck Creek, the future resulting from Scenario 3 generated the highest economic returns of any alternative. In this future, a large share of land was devoted to strip intercropping practices that were estimated to be highly profitable, in part because of the perennial grass crops they incorporated.

### *Environmental Measures*

The environmental measures calculated by i_EPIC for Walnut Creek were nitrate-N runoff, nitrate-N leaching, soil water erosion, and soil wind erosion (Table 5-2a). Data in Table 5-2 for nitrate-N export and sediment export from the watershed were obtained from Vaché et al. (2002), who used the SWAT model.

The corn–soybean rotation that dominates the baseline landscape for Walnut Creek and the Scenario 1 futures of both watersheds generates the largest environmental degradation, whereas grass hay causes the least. For example, in the Walnut Creek baseline, the average nitrate-N runoff per acre of cropland is 4.9 lb (5.5 kg/ha). Nitrate-N leaching is significantly higher, averaging 9.1 lb/acre (10.2 kg/ha). The grain oat–alfalfa enterprise contributes significantly less nitrate-N pollution to the watershed with 1.7 lb/acre (1.9 kg/ha) of nitrate-N runoff and 4.4 lb/acre (4.9 kg/ha) of nitrate-N leaching.

Corn and soybean production also cause the most soil erosion. Both water and wind erosion rates for this enterprise are above 3.19 short tons (st)/acre (8 metric tons [t]/ha). Water and wind erosion rates are below 0.80 st/acre (2 t/ha) with an oat–alfalfa rotation. Erosion rates and nitrate runoff and leaching from grass hay fields are negligible.

In Scenario 1, corn–soybean production is expanded at the expense of other land use. With the move to conservation tillage in the Scenario 1 future, however, all i_EPIC measures of environmental degradation except nitrate-N leaching were lower for Walnut Creek than in the baseline landscape. The adoption of conservation tillage reduces soil erosion significantly in Walnut Creek under Scenario 1 compared to the baseline landscape, even though more land is devoted to corn and soybean production (Table 5-2a). This result is consistent with some previous studies on the effect of conservation tillage on nitrate-N water pollution (e.g., Tyler and Thomas 1979; Thomas et al. 1981; Wu and Babcock 1999). For Buck Creek, the expansion of corn–soybean production at the expense of land formerly in pasture under Scenario 1 leads to increased nitrate export and relatively high sediment export rates (comparable to the baseline landscape; Table 5-2b).

Compared with other alternatives, Scenario 2 scored highest on the environmental measures of water quality, followed closely by Scenario 3. These results were consistent in both modeling approaches for both watersheds (Table 5-2a,b). The Scenario 2 futures produced the lowest total nitrate-N runoff and leaching and low levels of wind erosion, although for Walnut Creek watershed, i_EPIC estimated that Scenario 3 would generate slightly lower water erosion. Sediment export is lowest under Scenario 2 in both watersheds. These results are largely the consequence of converting cropland from corn and soybean production to perennial cover: rotational grazing, grass hay, and alfalfa. Surprisingly, average nitrate-N leaching is relatively high for some crop enterprises in Scenario 2. A large portion of the nitrate-N loss comes from the six-year crop rotation, in which the average nitrate-N leaching rate is 11.7 lb/acre (13.1 kg/ha) in the Walnut Creek watershed. Alfalfa also has a fairly high rate of nitrate-N leaching at 8.1 lb/acre (9.1 kg/ha). Very little nitrate-N runoff and water and wind erosion come from alfalfa and grass hay fields near the stream.

Scenario 3 incorporated the use of no-till practices as well as strip intercropping in the agricultural operations. The average nitrate-N runoff for this alternative was estimated to be 2.4 lb/acre (2.7 kg/ha) in the Walnut Creek watershed, less than that of the baseline landscape or the Scenario 1 alternative future, though higher than for Scenario 2. Using i_EPIC, the overall average nitrate-N leaching for Walnut Creek in this future, however, is 14.0 lb/acre (15.7 kg/ha). Under i_EPIC modeling assumptions, which do not account for nutrient uptake or transformation in depositional environments, the two nonorganic strip intercropping systems generate more nitrate-N leaching than traditional corn–soybean rotations.

The organic farming system generates a nitrate-N leaching rate of 8.3 lb/acre (9.3 kg/ha), which is approximately 7.7 lb (3.5 kg) less than that of the traditional two-year corn and soybean rotation. In the organic farming system, no artificial fertilizer is applied. Total water erosion was only 5,262.46 st (5,347 t), which is less than in any of the other scenarios. The corn–soybean rotation is the largest contributor to wind erosion in this scenario. Total wind erosion for the Walnut Creek watershed was estimated at 6,822.40 st (6,932 t), which is much lower than that in the baseline and production scenarios, but higher than Scenario 2.

Results from the SWAT model are similar to those obtained from i_EPIC, in that nitrate export and sediment export are slightly higher in the Scenario 3 future for both watersheds than under Scenario 2, but lower than under Scenario 1 (Table 5-2b). For landscapes dominated by corn and soybean row crops, such as the current landscape of Walnut Creek or the Scenario 1 future in Walnut Creek, the estimate of nitrate export obtained by summing the runoff and the leaching from i_EPIC was 93 percent of what the SWAT model estimated would be removed by streamflow from the watershed.

Results from SWAT and i_EPIC diverged, however, for both Scenarios 2 and 3. Both scenarios contain substantial amounts of perennial vegetation (either in the strips of native perennials in the strip intercropping and in the bioreserves; or in the rotational grazing, alfalfa, and hay fields adjacent to the fields of corn and soybeans), as well as riparian areas designed to provide depositional environments for nutrient uptake and denitrification. Because EPIC is a field-based model and does not account for nutrient uptake or transformation in depositional environments outside of the fields, we would expect that EPIC would overestimate runoff and leaching at the watershed scale, especially in the alternative futures of Scenarios 2 and 3, where many features of the landscape have been designed for nutrient uptake and transformation.

## Economic and Environmental Objectives: Complementary or Conflicting?

Much research has focused on the impact of agriculture on water quality (e.g., Hallberg 1989; Gren 1993; Kronvang et al. 1995; Gren et. al. 1997; Bystrom 1998), and on the impact of various tillage and conservation practices on both farm income and water quality. Concern has also been voiced about the fate of applied nutrients, particularly nitrate-N—the most commonly detected pollutant in groundwater—and their possible effects on water quality (Mueller et al.1995). Although some studies (e.g., Tyler and

Thomas 1979; Thomas et al. 1981; Wu and Babcock 1999) found that conservation tillage increases nitrate-N leaching, perhaps as a result of an increase in soil porosity created by increased organic matter and microbial activity (Jacobs and Timmons 1974; Blevins et al. 1977, 1983; McMahon and Thomas 1976), others observed no difference or less leaching with conservation tillage (Kitur et al. 1984; Kanwar et al. 1985). Conservation tillage thus has the potential to increase or decrease nitrate-N leaching, depending on nutrient additions, soil quality, and weather conditions (Gilliam and Hoyt 1987). Different results emerge from two different modeling approaches (i_EPIC and SWAT) to estimate nitrate runoff, nitrate leaching, and nitrate export from these cropping systems. The differing results highlight the importance of understanding the influence of cropping systems and agricultural practices, as well as the flow of water through various settings on the movement of nitrate through the system.

Ranking of landscapes from most to least preferred is not strictly an economic or environmental proposition. The data presented in Table 5-2 illustrate the degree to which economic and environmental objectives can be achieved together or the degree to which trade-offs are involved. For example, compared to the baseline landscape, Scenario 1 improves RTL and some environmental measures. In Scenario 2, improved environmental quality comes at an economic cost.

Not all trade-offs resulting from changes in land use are between economics and the environment. Sometimes there may be difficult tradeoffs between different components of environmental quality. For example, the switch from conventional to conservation tillage improves soil erosion and nitrate-N runoff measures but may increase nitrate-N leaching, as in Walnut Creek watershed under Scenario 1. In the alternative futures for Walnut Creek under Scenario 3, we find less water erosion than under Scenario 2, but higher wind erosion. Each alternative generates some negative consequences along with the gains. Changes in agricultural practices should be targeted to specific environmental goals, and the trade-offs accompanying each choice should be carefully considered for their impacts on economic and environmental goals.

## Evaluating Our Results

Our modeled results indicate that considering the entire Walnut Creek watershed, the future landscape under Scenario 1 would produce the highest RTL among the four alternatives evaluated in this study. Under Scenario 1, total harvested acres would increase and production costs would decrease as the result of adopting no-till practices. Considering the entire Buck Creek watershed, however, the Scenario 3 future generated the highest returns because of the profitability of strip intercropping with native prairie strips. The Scenario 2 future produces the lowest RTL of the four alternatives, but scores highest on the environmental measures related to water quality. The alternative futures that result under Scenario 3 generate lower erosion; however, they also lead to the highest rate of nitrate-N leaching according to the i_EPIC model. We should note, though, that under conditions in which subsurface flow is expected to be captured by tile drainage systems and routed through landscape features designed to reduce nutrient export from the watershed (as modeled by Vaché et al. [2002] using the SWAT model), nitrate leaching into groundwater may be much lower under Scenarios 2 and 3. In the designs for Scenarios 2 and 3, the flow from tile drains is directed into constructed wetlands before it flows into the creek, allowing nitrate removal by plant uptake and denitrification.

Boody et al. (2005) conducted a similar study of alternative futures for two agricultural watersheds in Minnesota. In their project, two local producers assisted both in developing scenarios and in documenting farm-specific, detailed descriptions of agricultural enterprises. The study team used net return to management as the economic endpoint (net farm income estimated as a function of output of crop and livestock products, based on five-year weighted average real output prices for Minnesota and cost of production in time and labor). Producer surveys were used to calculate production costs. Average outputs of agricultural products were used for each production system, and adjusted to reflect varying soil quality. Their economic analyses were based on outputs from areas in each production system summed for the watershed. As in our study, they did not include government commodity payments in their estimate of farm income (except for an area-weighted apportionment of county dairy payments); however, they did provide an estimate of potential government commodity payments for each of their futures in each watershed. They used the ADAPT (Agriculture Drainage and Pesticide Transport) model (Gowda et al. 1999) to estimate water quality impacts of the varying land use and agricultural practices in their scenarios. Their endpoints were N, phosphorous (P), and sediment delivery to the mouth of the streams in their study watersheds, estimated by aggregating edge-of-field estimates across the study area, an approach comparable to that used in the EPIC model. Their results were consistent with those from our study, in that substantial changes in agricultural practices were required to yield substantial improvements in water quality, and grain production had a higher profitability than livestock production (outside confinement operations) given current patterns of pricing for these commodities.

## Implications for Agricultural Policy

The agricultural practices modeled for the baseline landscape result in the largest amount of soil erosion and nitrate-N runoff of the four scenarios. These results suggest that substantial improvements over the baseline landscape are possible.

The finding that different results can emerge from using two different water quality modeling approaches highlights the need for research on the conditions under which nitrate leaching may be a significant problem. Research is also needed on the effectiveness of landscape features such as engineered wetlands and riparian buffers designed to remove nutrients from runoff and tile drainage flow.

In summary, the extent to which economic and environmental objectives can be achieved together will depend heavily on costs of fuel, labor, and machinery, along with the productivity and continued profitability of environmentally targeted farming practices. Careful planning and management of nutrient applications during transitions from one cropping system to another may be needed to avoid unintended consequences (for example, increased nitrate leaching into subsurface water) that could result in the short term from practices intended to improve soil and water quality over the long term.

## Acknowledgments

This study was a part of a project funded by the U.S. Environmental Protection Agency (U. S. EPA) Star Grants program (Water and Watersheds, grant # R-825335-01). We thank Kellie Vaché for help with the GIS data, as well as other members of the project. We also thank seminar participants at the Heartland Environmental Resource Economics Conference in Ames, Iowa, in September 1999, and anonymous referees for helpful comments.

## Chapter 6

# Farmers' Perceptions

## Joan Iverson Nassauer, Robert C. Corry, and Jennifer A. Dowdell

How might farmers respond to the alternative futures described in Chapter 4? To begin to anticipate their responses, we conducted in-depth interviews with Iowa farmers, visiting them in their homes or businesses. From these interviews, we learned that farmers recognize the value of alternative agricultural policies, practices, and technologies that improve environmental quality—even if future agricultural landscapes look very different from those of today and suggest new management challenges. Based on their perceptions, we were also able to map farmers' relative preference for different land covers in the two study watersheds (Walnut Creek and Buck Creek; Color Figures 44 and 45). This allowed us to compare overall preference for the three alternative scenarios and the baseline landscape from the perspective of Iowa farmers. The comparison showed that in both the flat, productive landscape of Walnut Creek (Color Figure 6) and the rolling, more erosive landscape of Buck Creek (Color Figure 10), farmers found Scenario 3 (Color Figure 18), which emphasized enhanced biodiversity, to be best for the future of the people of Iowa. Scenario 2 (Color Figure 17), which emphasized enhanced water quality, ranked a close second.

Water quality, which affects both hypoxia of the Gulf of Mexico (Chapter 2) and quality of life in the Corn Belt (Chapter 3), improved dramatically under either Scenario 2 or Scenario 3 (Chapter 7; Color Figure 16). The Corn Belt integrated assessment (Chapter 14) further indicates that Scenario 3 performed well by nearly every measure including biodiversity measures and economic return to land (RTL). This suggests that policy that promotes innovative land covers and practices that accomplish multiple environmental and societal functions in Corn Belt agricultural landscapes could be acceptable to farmers.

Interview results also suggest that what farmers want for Corn Belt landscapes may not be all that different from what the larger public wants. Both groups will influence the future of agricultural landscapes through their choices as constituents and consumers, and both groups live with the environmental effects of agricultural practices. Farmers and the public, though, have often been understood to hold conflicting values about agriculture (Merrigan 1997; Walter 1997; Beedell and Rehman 1999). Farmers are sometimes assumed to be less concerned and less knowledgeable about agriculture's environmental effects, and the public is assumed to be more concerned about food safety and the environmental degradation associated with agriculture (McHenry 1997; Paolisso and Maloney 2000).

In fact, farmers have generally been found to be highly knowledgeable about the environmental effects of agriculture, including its contributions to water pollution, and to value practices that promote long-term productivity of agricultural land, healthy rural communities, and good stewardship (Allen and Bernhardt 1995; American Farm Bureau 1998; Paolisso and Maloney 2000; Dutcher et al. 2004; Hudson et al. 2005; Urban 2005). At the same time, farmers tend to be willing to use new technology (Abaidoo and Dickinson 2002), and the appearance of large-scale industrial agricultural landscapes may contribute to public misimpressions of farmers' stewardship values (Nassauer 1997; Jackson and Jackson 2002). The contradiction may not be between farmers' values and public values. Instead, farmers' values may not always be apparent in the appearance of contemporary agricultural landscapes. Farmers could benefit from future policies that anticipate that agricultural landscapes are the most visible products of farm management choices. Such policies might promote not only ecological benefits and technologies to produce commodities, but also landscape characteristics that both farmers and the broader public recognize as valuable.

In the United States, unlike other member nations of the Organisation for Economic Co-operation and Development (OECD), little attention has been paid to farmers' perceptions of agricultural landscapes (Dramstad and Sogge 2003). Yet Corn Belt farmers know the importance of driving around the farm neighborhood to see crop conditions and observe the practices that other farmers are using. Farmers' perceptions of agricultural landscapes suggest that economic benefits alone do not account for their landscape preferences, willingness to adopt sustainable agricultural practices, or attitudes toward environmental issues (Nassauer and Westmacott 1987; Nassauer 1988; Schaumann 1988; Nassauer 1989; Napier and Brown 1993, Milham 1994; Willits and Luloff 1995; Kline and Wichelns 1996; McCann et al. 1997; McHenry 1997; Ryan 1998; Traore et al. 1998; Willock et al. 1999; Beedell and Rehman 2000; Oglethorpe et al. 2000).

Perception is not just a matter of impressions or superficial amenities. What people see informs their attitudes, and their attitudes affect their behavior (Ervin and Ervin 1982; Featherstone and Goodwin 1993; Traore et al. 1998; Willock et al. 1999). Farmers' perceptions of their neighbors' attitudes, their family's attitudes, the practices used by other farmers, erosion and water quality on their farms, visual aesthetics of their land, recreational values, and risks to their health can all affect their management choices (Bultena and Hoiberg 1983; Gould et al. 1989; Napier and Brown 1993; Vogel 1996; Salamon et al. 1997; van den Berg et al. 1998; Beedell and Rehman 1999; Lichtenberg and Zimmerman 1999; Napier et al. 2000; Ryan et al. 2003; Carolan 2004; Urban 2005). Even how they think the public perceives *them* may influence their choices (McHenry 1997; American Farm Bureau 1998). Societal perceptions of the broad public benefits of federal agricultural policy could be affected, in part, by perceptions of changing agricultural landscapes.

The sections that follow describe our methods for measuring farmers' perceptions in this study.

## Representing Landscape Futures

Based on the alternative futures for Buck Creek and Walnut Creek watersheds (Chapter 4; Color Figures 38–43), we developed digital imaging simulations of particular views of future landscapes as seen from the ground (Color Figures 13–18) and from a low-level flight (Color Figures 4–12). We also selected photographs that showed current features of Corn Belt landscapes that could exist in 2025 (Color Figures 19–27). The photographs showed features such as strip intercropping, pasture, constructed wetlands, a restored stream corridor, a prairie remnant, a concentrated animal feeding operation (CAFO), and even large-lot residential development in an agricultural area. Then we used 8 × 10 inch color prints of both the simulations and photographs during our in-depth interviews of Iowa farmers.

## Selecting Farmers

We interviewed 32 Iowa farmers who lived in townships with similar soils and in the same counties as the Iowa study watersheds. To avoid bias that might be introduced by ownership or neighbor relationships, we selected townships outside the study watersheds themselves. We targeted farmers who represented a range of enterprises and farm sizes (Table 6-1). Mean farm size was 360 acres (146 ha) for each county, and median farm size was 210 acres (85 ha) for Poweshiek County and 180 acres (73 ha) for Story County (USDA 1997). We also targeted farmers who had demonstrated their ability to be innovative with land cover; about half the interview participants had land currently or previously enrolled in the Conservation Reserve Program (CRP), compared with about 26 percent of all Iowa farmers participating in the CRP in 1997 (USDA FSA 2002).

Once we selected the townships for interviews, Iowa State University county extension agents supplied us with the names of all 83 farmers living in the zip codes of those townships. When contacted by phone, 27 agreed to be interviewed, and 24 of those were able to keep their interview appointments. To achieve our desired range in size of area farmed, we augmented this sample by asking the original 24 interviewees to name nearby farmers who farmed large amounts of land. From these names, we contacted 16 and interviewed 8 additional farmers. All but 3 of the farmers we interviewed drew at least 75 percent of their family income from farming; 9 earned all their income from farming. All but 4 were men. Eighteen were 50 or younger and 14 were older than 50.

## Conducting Interviews and Gathering Data

Pairs of trained interviewers spoke to the farmers in their homes (except two who were interviewed in a local restaurant and in a local business, respectively). The structured

**Table 6-1.**  Characteristics of Farmers Interviewed

| *Farm sizes* | *Story County*[a] | | *Poweshiek County*[b] | |
|---|---|---|---|---|
| | Hay producer | No hay | Hay producer | No hay |
| 80 acres (32 ha) or fewer | 0 | 2 | 1 | 0 |
| 81–640 acres (33–259 ha) | 3 | 8 | 7 | 4 |
| More than 640 acres (259 ha) | 0 | 4 | 2 | 1 |

[a] Twenty-eight percent of farms produce hay.
[b] Fifty-two percent of farms produce hay.

interview format had three parts. First, while viewing 11 color prints of future land-scapes and their features (Color Figures 16 and 18–27), the farmers used their own words to describe what they noticed, what they found attractive or unattractive, and what they thought would be advantageous or disadvantageous for farming in each landscape. Interviewers were trained to furnish no information about the landscapes in the prints or the alternative scenarios before farmers gave their descriptions. Overall, the open-ended questions to the 32 farmers yielded 352 landscape descriptions, which were then subjected to a content analysis (Weber 1990).

Next, the interviewers provided information about the alternative policy scenarios while showing the farmers 9 additional 8 × 10 inch color prints of simulations of the futures described in Chapter 4 (Color Figures 4–12). The farmers were asked to con-sider what landscapes would be best for the people of Iowa in 25 years, assuming that because of farm policy programs, net farm income would be the same for each of the alternatives. Using a Q-sort forced rating method (Addams 2000; Fairweather and Swaf-field 2000), each farmer sorted the original 11 images of farming practices, enterprises, or land cover types into three piles, ranging from what was "best for the future" (3 im-ages), through what was "next best for the future" (5 images), to what would be "least good for the future" (3 images). Then they repeated the process with all 20 images, including the first 11 prints of landscape futures and their features, into five piles of 3–4–6–4–3 images. These piles ranged from what was best for the future (3 images rated "5"), through what was next best for the future (4 images rated "4"), to what would be least good for the future of the people of Iowa (3 images rated "1"). Overall, the second Q-sort yielded 640 ratings of the 20 landscape images.

## Content Analysis Results

The content analysis of the farmers' 352 descriptions of the first 11 images distin-guished several themes (Weber 1990). Overall, the 2,668 phrases that the farmers used in their descriptions of these images suggest that farmers are highly attuned to the ap-parent stewardship of the agricultural landscape, and that they perceive the landscape not only in terms of its enterprises and potential productivity but also in terms of its amenity values. Because farmers received no information about the policy scenarios before describing the images, their frequent mention of stewardship terms tends to validate their subsequent high ratings of Scenario 3, which had a leading goal of en-hancing biodiversity.

Table 6-2 presents the concepts farmers used most frequently to describe the landscapes. By far the largest number of phrases relate to good stewardship. These include specific conservation practices that were noticed (e.g., buffer strips, crop resi-due, grassed waterways, and nutrient management) as well as evaluative phrases (e.g., "good job of conservation farming," "that's planted the right way for the hill," "well suited to the land capability," and "well farmed for what's there"). In answer to ques-tions about what they found unattractive or a disadvantage for farming, farmers fre-quently used phrases related to poor stewardship (e.g., "[this farmer is] a user not a caretaker," "high use of fertilizers," "no visible crop rotation," "too intense," and "not on the contour").

Another large number of phrases related to enterprises. Farmers frequently noticed pasture, crops, and trees, and all but one farmer discussed the CAFO and associated concerns about offensive odors (Color Figure 26). Apparent soil quality and productiv-ity (and visual cues related to soil quality, such as topography, woodlands, wetness, or

**Table 6-2.** Content Analysis Results[a]

| What farmers noticed | Number of times farmers used this concept | What farmers found attractive or an advantage | Number of times farmers used this concept | What farmers found unattractive or a disadvantage | Number of times farmers used this concept |
| --- | --- | --- | --- | --- | --- |
| Good stewardship | 155 | Good stewardship | 380 | Poor soil quality | 118 |
| Good pasture | 96 | Attractive landscape | 83 | Poor stewardship | 53 |
| Crops | 80 | Wildlife or habitat | 70 | Looks dead or dry | 49 |
| Trees | 65 | Good pasture | 58 | | |
| Rolling land | 49 | Trees | 57 | | |
| Poor soil quality | 43 | Looks productive | 51 | | |
| | | Looks like a good return | 49 | | |
| | | Cattle | 44 | | |

[a] Concepts most frequently used (n > 40) by farmers to describe landscapes shown in Color Figures 4–12, 16, and 18–27.

dryness) figured heavily in the farmers' perceptions. Finally, farmers also commented on amenity landscape characteristics separate from their value as an enterprise (e.g., the attractiveness of the landscape, wildlife or habitat, and the attractiveness of cattle).

## Rating of Land Cover Types and Practices

Using the Q-sort data, we found the mean rating of each image to be very reliable; ratings of the same images in the first and second Q-sort ratings were significantly correlated (Pearson's $r = 0.732$, two-tailed significance, $p < 0.000$). In addition, image format (ground level versus aerial view) had no significant effect on ratings ($t = 0.157$, 18 df, $p < 0.88$).

Farmers' descriptions of the images before they were given any information about the images or policy scenarios were highly consistent with their ratings of the same images after they were given scenario information. For example, interviewers did not describe Color Figure 20 as a prairie or natural area, but farmers frequently volunteered that this image showed prairie or native plants. Images that were frequently ($n \geq 15$) described as wildlife and habitat had high ratings with one exception—Color Figure 25, the small wetland, was recognized for its wildlife and habitat value, but it was also described as weedy, and its mean rating was lower than the other habitat images. Images depicting land covers that would be part of Scenario 3 were consistently rated higher than other images. The farmers tended to describe these images as being productive, showing good stewardship, and sometimes, as benefiting wildlife and habitat.

Farmers in Story County, the location of flat Walnut Creek watershed, and Poweshiek County, the location of rolling Buck Creek watershed, had very similar perceptions about what would be best for the future of the people of Iowa. They differed significantly ($p < 0.05$) only in their ratings of flat Walnut Creek watershed in Scenario 3 (Color Figure 9; Story County, $m = 4.18$; Poweshiek County, $m = 3.33$) and of the CAFO with a plowed field (Color Figure 26; Story, $m = 2.18$; Poweshiek, $m = 1.47$). Farmers from the more rolling area gave lower ratings to both images. As they viewed the flat Walnut Creek watershed managed with strip intercropping in Scenario 3 (Color Figure 9), some farmers from rolling Poweshiek County remarked that the good flat land was too productive to be managed with strips, expressing that, in their opinion, it was not the best use of that land. As they examined the CAFO (Color Figure 26), several farmers in Poweshiek County related vivid memories of a spill from a local hog confinement facility lagoon into a local stream. This helps to explain the differences between the counties.

More remarkable are overwhelming similarities in perceptions of farmers from these two very different Corn Belt terrains. Overall, the landscape images that farmers perceived as best for the future of Iowa (those rated 3.60 or higher) were Color Figures 9 and 18–21 (Table 6-3). All but one of these images show land covers under Scenario 3; it (Color Figure 21) shows the current landscape. Surprisingly, Corn Belt farmers perceived the prairie as being very good for the future of the people of Iowa. Content analysis showed that they recognized the prairie as being attractive, having native plants, having flowers, and showing good conservation. In addition, their high rating of strip intercropping images suggests that they recognize environmental benefits, even though they were concerned that these practices could be "difficult to manage with big equipment." Farmers perceived the practices portrayed in Color Figures 5, 6, 8, 12, 16, and 22–24, which show a variety of different land covers in all of the scenarios (as well as the present), as very good—rated between 3.01 and 3.59—for the future of the

**Table 6-3.** Farmers' Ratings and Descriptions of Alternative Scenarios

| Scenarios depicted | Image content and Color Figure number | Mean rating n = 32 | Standard deviation[a] | Descriptors |
|---|---|---|---|---|
| 3 | Walnut Creek watershed scenario emphasis on enhanced biodiversity; Color Figure 9 | 3.78 | 1.49 | Not included in part 1 of interview protocol |
| 3 | Strip intercropping with perennial prairie strips; Color Figure 18 | 3.75 | 1.13 | Good stewardship, difficult farm management, good yield or productive |
| 3 | Strip intercropping of corn, soybeans, and oats; Color Figure 19 | 3.75 | 1.44 | Difficult farm management, good stewardship, good yield or productive |
| Baseline, 3 | Prairie remnant; Color Figure 20 | 3.66 | 0.91 | Return poor, return good, good stewardship, prairie and native plants, attractive wildlife and habitat |
| Baseline | Current land cover types in upper reaches of Buck Creek watershed; Color Figure 21 | 3.62 | 0.61 | Good stewardship, trees attractive |
| 2 | Buck Creek watershed scenario emphasis on improved water quality; Color Figure 12 | 3.56 | 1.31 | Not included in part 1 of interview protocol |
| 2 | Walnut Creek watershed scenario emphasis on improved water quality; Color Figure 8 | 3.56 | 1.12 | Not included in part 1 of interview protocol |
| 2,3 | Riparian corridor restoration; Color Figure 22 | 3.47 | 0.81 | Good stewardship, stream, attractive wildlife and habitat, attractive landscape, good water quality, flooding disadvantage |
| 3 | Buck Creek scenario emphasis enhanced biodiversity; Color Figure 5 | 3.28 | 1.52 | Not included in part 1 of interview protocol |
| Baseline | Walnut Creek current conditions; Color Figure 6 | 3.22 | 0.98 | Not included in part 1 of interview protocol |
| Baseline, 2 | Pasture; Color Figure 23 | 3.16 | 1.13 | Cattle good, pasture, good stewardship, poor soil quality, dead or dry, unattractive |
| Baseline, 2 | Rolling fields of row crops; Color Figure 24 | 3.12 | 1.17 | Good stewardship, poor soil quality, pasture |
| Baseline, 1, 2, and 3 | Buck Creek scenario, emphasis on production. Rolling corn field; Color Figure 16 | 3.06 | 0.94 | Good stewardship, difficult farm management, crops, poor soil quality |

**Table 6-3.**   Farmers' Ratings and Descriptions of Alternative Scenarios (continued)

| Scenarios depicted | Image content and Color Figure number | Mean rating n = 32 | Standard deviation | Descriptors |
|---|---|---|---|---|
| **Baseline, 1, 2, and 3** | Small wetland; Color Figure 25 | 2.66 | 0.91 | Attractive wildlife and habitat, weedy, good pasture, good stewardship |
| 1 | Walnut Creek watershed scenario, emphasis on commodity production; Color Figure 13 | 2.66 | 2.04 | Not included in part 1 of interview protocol |
| **Baseline** | Buck Creek current condition; Color Figure 10 | 2.53 | 1.19 | Not included in part 1 of interview protocol |
| **Baseline** | Buck Creek current conditions, eroding pasture; Color Figure 4 | 2.10 | 1.15 | Not included in part 1 of interview protocol |
| **Baseline, 1, 3** | CAFO; Color Figure 26 | 1.84 | 1.01 | Good stewardship, big business disadvantage, difficult farm management, big animal confinement, poor odor |
| **None** | Residential development; Color Figure 27 | 1.84 | 0.57 | Unattractive, exurban, poor soil quality, poor crop health. |
| 1 | Buck Creek watershed, scenario emphasis on commodity production; Color Figure 11 | 1.47 | 0.81 | Not included in part 1 of interview protocol |

[a] Standard deviation of sample distribution. Smaller standard deviations indicate less variation in the ratings of the farmers sampled.

people of Iowa. In the content analysis, both the stream buffer (Color Figure 22) and the pasture (Color Figure 23) were described as showing good stewardship.

Color Figures 4, 7, 10, 11, and 25–27 contain what farmers perceived as least good (rated 3.00 or lower) for the future. Two of these specifically depict Scenario 1 (Color Figures 7 and 11), which maximized commodity production. Three of them show aerial views of rolling landscapes that farmers tended to describe as eroding under cultivation (Color Figures 4, 10, and 11). Remaining images show controversial land uses. For example, a CAFO (Color Figure 26) was described in the content analysis as a good example of animal confinement and an instance of big business. In addition, farmers mentioned that the manure would be difficult to manage and would produce odor. A small wetland in a field (Color Figure 25) was described in the content analysis as weedy, wasteland, poorly drained, and wildlife habitat. Exurban residential development among farms (Color Figure 27) was described in the content analysis as being non-farm development and having poor-looking crops.

Overall, our results suggest that farmers are highly knowledgeable about the environmental implications of various practices and land covers; that they see the value of innovation to achieve ecological benefits; and that they personally value ecologically ben-eficial land covers, land uses, and practices as part of the larger agricultural landscape.

## Mapping Farmer Preferences for the Alternative Futures

To map and quantitatively measure the farmers' perceptions of the alternative futures compared with the baseline landscape for each watershed, we selected subsets of images that depicted the land covers and practices for each particular scenario. Table 6-3 identifies images selected to represent each scenario. On average, the land covers in Scenario 1 rated lowest and the land covers under Scenario 3 rated highest. Next, we translated the farmers' mean ratings of different land covers to map their preferences for the alternative landscape futures for both study watersheds (Color Figures 38–43). To do this, we used the land cover geographic information systems (GIS) databases for each watershed and each scenario shown in Color Figures 6, 10, and 38–43, and, based on the following heuristic, assigned farmers' perception ratings of the images to each of the 87 land cover classes in the 4 land cover databases (for the baseline landscape and the three alternative scenarios) for each of the two watersheds ($n = 8$). Our heuristic applied the following rules in a hierarchy that favored the most simple rule applicable to each land cover class. To help us determine how well a given image represented a land cover class, we referred to the texts of the farmer interviews. For example, if farmers had frequently spoken about stream buffers when they viewed an image, we judged that image to be a good representation of stream buffers. The rules, ranging from most to least simple, are as follows:

1. Apply the mean rating of one image (100 percent) if it adequately represents the land cover class:

$$LC_x = \bar{x} \text{ Image}_a.$$

2. Apply weighted mean ratings of two images (70/30 percent) if one image is more representative than the other, but each is representative:

$$LC_x = 0.7 \ \bar{x} \text{ Image}_a + 0.3 \ \bar{x} \text{ Image}_b.$$

3. Apply equally weighted mean ratings of two images if both are equally representative:

$$LC_x = 0.5 \ \bar{x} \text{ Image}_a + 0.5 \ \bar{x} \text{ Image}_b.$$

4. Apply weighted mean ratings of three images if one image is more representative but all are representative:

$$LC_x = 0.5 \ \bar{x} \text{ Image}_a + 0.25 \ \bar{x} \text{ Image}_b + 0.25 \ \bar{x} \text{ Image}_c.$$

Using this heuristic to infer land cover class preference ratings, we mapped farmer preference of each scenario applied to each watershed (Color Figures 44 and 45). Then, we calculated an area-weighted farmer preference score for each scenario for each watershed, where the overall farmer landscape preference for each scenario in a given watershed is *LP*, the area of a given land cover class is *a*, and the land cover class preference rating is *p*:

$$LP = \frac{\sum_{i=1}^{87}(a \times p)}{\sum_{i=1}^{87} a}$$

Results of the area-weighted preference analysis for each watershed (Color Figures 44 and 45) show that the ranking of the scenarios is the same for flat Walnut Creek watershed and rolling Buck Creek watershed. In each case, Scenario 3 was most preferred, followed by Scenario 2, the baseline landscape, and Scenario 1. Results also show that the largest difference between Walnut Creek and Buck Creek watersheds is in preference for Scenario 1, which emphasized increased commodity production as the highest priority. This is consistent with farmers' interview observations that the Walnut Creek watershed appeared to be more suitable for crop production, and the Buck Creek watershed appeared to be more vulnerable to erosion. In both watersheds, the scenarios rated by Iowa farmers as best for the future of the people of Iowa were those that improved water quality and increased biodiversity.

## Consistent Themes from the Study

A highly consistent picture emerges from these multiple measures and analyses. Characteristics that farmers described as attractive or advantageous before we told them about the alternative policy scenarios also emerged as what farmers rated as best for the future of the people of Iowa after they had been told about the scenarios. And the descriptions that they used suggest that the farmers are highly attuned to their role as land stewards, as well as to the land's values for both production and amenity. Consistent with related studies, these analyses suggest that the Iowa farmers we interviewed are knowledgeable about environmental characteristics of agricultural landscapes, that they are open to new policies and new technological innovations that would allow them to manage their farms for greater environmental benefits, and that they place a high value on stewardship by farmers (Paolisso and Maloney 2000; Ryan et al. 2003; Dutcher et al. 2004; Urban 2005). If their own incomes would not suffer as a result, the farmers we interviewed tended to see future landscapes that exhibit long-term stewardship as attractive or advantageous *and* as being best for society—even when that meant changing their practices.

Methodologically, this project demonstrates that measuring farmers' perceptions of and preferences for whole landscapes, as represented by realistic images, can reveal different information than investigating farmers' attitudes and beliefs about more abstract concepts or specific practices. Eliciting farmers' perceptions of landscapes that they are actually seeing requires them to apply their underlying attitudes and beliefs to the complex landscape contexts in which they make farming decisions. Consequently, their responses may reflect a more complete set of values for their land. These values could be stated, for example, as productive land, viable habitat, desirable home, good real estate investment, and pleasant aesthetic experience (Schrader 1995; van den Berg et al. 1998; Napier et al. 2000; Schoon and te Grotenhuis 2000; Napier and Bridges 2002; Ryan et al. 2003; Urban 2005). Although our findings are consistent with some previous investigations of farmers' attitudes, our study was different from many in that farmers actually saw the landscapes they described and rated, including some that displayed unfamiliar, innovative practices. Our method may help to anticipate acceptance of future policies that introduce innovation. Cultural values have influenced many environmental policy decisions since 1970. Going further to explicitly map local values as we did here may help to advance policy implementation.

## Implications for Agricultural Policy

In the eyes of the nonfarming public, agricultural landscapes create an impression of farmers' apparent environmental values. But currently, such perceptions can be misleading—giving the public the impression that farmers are unconcerned about broader environmental benefits. Results of our study of Iowa farmers suggest that, presented with an income-neutral, ecologically beneficial policy alternative for managing the landscape for long-term agricultural production, farmers tend to choose what is best for the environment. Farmers place a higher value on innovation for ecological benefits than policymakers or the larger public might guess.

Policies of many of America's global trade partners use public investment to support a wide range of societal and environmental benefits produced by their rural landscapes (Westmacott and Worthington 1984; McHenry 1997; Wilson 1997; Swaffield 1998; van den Berg et al. 1998; Beedell and Rehman 1999; Hamblin; Menzies 2000; Wragg 2000; European Communities 2001). Future U.S. agricultural policy could protect and support characteristics of the agricultural landscape that are seen as most valuable by farmers and by the larger public, particularly characteristics that are immediately recognizable for their environmental benefits. Policy that forges links between a broad range of agricultural landscape benefits and public expectations may lead to future agricultural landscapes that also more closely resemble farmers' own preferences for the future.

## Acknowledgments

We are grateful to the farmers who shared their valuable insights during the interviews, each of which lasted several hours. We also extend thanks for the essential work of our colleagues, Kathryn Freemark Lindsay and Marilee Sundt, who participated in conducting the interviews. We thank the U.S. Environmental Protection Agency (U.S. EPA) STAR grants program (Water and Watersheds, grant #R-825335-01) for funding. Although the research cited here was funded by the U.S. EPA, the conclusions and opinions presented here are solely those of the authors and are not necessarily the views of the agency.

*Chapter 7*

# Water Quality

### Kellie B. Vaché, Joseph M. Eilers, and Mary V. Santelmann

Despite advances in our understanding of agriculturally derived nonpoint source (NPS) pollution and the development of best management practices (BMPs) intended to reduce NPS pollution (USDA SCS 1994), aquatic ecosystems linked to agricultural regions in the United States continue to receive high loadings of agricultural pollutants. The mechanisms associated with these increased loadings are complex. It is well established, however, that they result from a combination of increased inputs of nutrients through soil amendments and fertilization and a decreased potential for nutrient immobilization and loss through improved land surface drainage related to wetland conversion, tile drainage, and channelization (Menzel et al. 1984; Mitsch et al. 2001). The effects of these human-induced changes are relevant on a wide variety of scales, from small watersheds (Becher et al. 2000; Schilling and Thompson 2000) to large basins such as the Mississippi River Basin (MRB; see Chapters 2 and 15). These effects extend into coastal waters such as the Gulf of Mexico, where nutrient-rich discharge from the Mississippi River significantly increases the potential for eutrophication and oxygen depletion. Many of the most effective remediation strategies for reducing NPS pollution, including improved nutrient management, off-field buffers, crop selection and rotation, and conservation set-asides, are implemented at the farm and local watershed scale. The cumulative effect of implementing the locally appropriate sets of these alternative strategies (see Brezonik et al. 1999), combined with basin-scale land use and management practices such as extensive restoration of riverine riparian wetlands and floodwater management structures (Chapter 15), could ultimately lead to the level of reductions in nutrient loadings necessary to substantially reduce eutrophication rates in the Mississippi River and the Gulf. For this reason, small watershed studies have significant potential to inform decisionmaking and planning across much larger areas, including the MRB and the Gulf of Mexico.

In this chapter, we focus on presenting the use of simulation modeling of small watersheds to quantify the potential effects of a suite of targeted land use and management changes on water quality. The changes we evaluate are presented in Chapter 4 as a set of three alternative scenarios applied in two watersheds, Buck Creek and Walnut Creek, Iowa. We modeled potential effects of the land cover patterns and management regimes described for each scenario compared with the baseline landscape in each watershed using the Soil and Water Assessment Tool (SWAT; Arnold et al. 1995), calibrated to current conditions in each watershed.

Water quality conditions over the model calibration period in Walnut Creek have been characterized through analysis of water quality data collected monthly since 1990 for Walnut Creek watershed in Story and Boone counties (Cambardella et al. 1999; Hatfield et al. 1999; Jaynes et al. 1999). We characterized water quality conditions in Buck Creek watershed (Poweshiek County) as part of this research (Shoup 1999). The model was also applied to historical land cover (Color Figures 31 and 35) to estimate past discharge and surface water chemistry. This allowed us to estimate the upper bound for water quality improvements that may be attainable in the region. Our objectives are to evaluate the potential effects of agricultural land use and management practices on water quality at the watershed scale, to present quantitative estimates of how landscape and management changes might affect water quality, and to discuss the implications of these results for agricultural policy.

## Baseline Landscapes and Future Scenarios

Tables 7-1 and 7-2 summarize the land cover data for the baseline landscapes of Walnut and Buck Creek watersheds, as well as for the past and future scenarios. Vaché et al. (2002) give assumptions about the agricultural practices (e.g., the amount and timing of fertilizer and tillage practices) used in each scenario.

In Scenario 1 priority is given to producing agricultural commodities while employing current BMPs (e.g., precision agriculture, no-till cultivation) as well as narrow riparian buffers to protect water quality. Scenario 2 includes numerous management practices designed to address water quality concerns (49- to 98-ft [15- to 30-m] riparian buffers, no-till cultivation, strip intercropping, and alfalfa fields or fenced pasture adjacent to stream buffers). Scenario 3 assumes broad-based public support for restoring native biodiversity, and many practices that would enhance biodiversity could also enhance water quality. Under Scenario 3, monocultures of corn and soybean rotations are replaced by strip intercropping that incorporates a strip of native perennials in fields of corn and soybeans. In addition, riparian buffer width doubles compared to Scenario 2. Production of organic crops increases and large areas are set aside as habitat reserves (Chapter 4).

Historical land cover designations were generated using a detailed (1:12000) regional soils database (ISPAID 2004), in which the designations are based on characteristics of soils formed under prairie and forest vegetation. Areas defined as prairie were subdivided into upland prairie, ephemeral wetland, seasonal wetland, semipermanent wetland, permanent wetland, and pond according to the relationships between soil types and associated wetland types described in Galatowitsch and van der Valk (1994).

Chapter 4 describes the study watersheds and alternative future scenarios. Methods used for collecting stream flow and water quality data used to calibrate the SWAT model, along with descriptions of model parameterization and calibration, can be found in Vaché et al. (2002). We summarize these methods in the sections that follow.

## Obtaining Water Quality Data

Walnut Creek is a study site in the U.S. Department of Agriculture (USDA) Management Systems Evaluation Area (MSEA) project. Five sites within the watershed have

**Table 7-1.** Land Cover Change in Walnut Creek

| Land cover | Baseline Area, (ha) | Scenario 1 | | | Scenario 2 | | | Scenario 3 | | |
|---|---|---|---|---|---|---|---|---|---|---|
| | | Area, (ha) | Change, (ha) | Change, % | Area, (ha) | Change, (ha) | Change, % | Area, (ha) | Change, (ha) | Change, % |
| Woodland closed | 66.4 | 61.9 | –4.5 | (–6.9) | 21.4 | –45.0 | (–67.7) | 126.4 | 59.9 | (90.2) |
| Woodland open | 114.2 | 69.4 | –44.8 | (–39.2) | 82.8 | –31.4 | (–27.5) | 109.1 | –5.0 | (–4.5) |
| Savannah | 41.3 | 24.3 | –16.9 | (–41.1) | 39.8 | –1.4 | (–3.5) | 34.4 | –6.8 | (–16.6) |
| Corn/soybeans | 4190.5 | 4510.4 | 319.8 | (7.6) | 2882.3 | –1308.2 | (–31.2) | 1725.1 | –2465.4 | (–58.8) |
| Perennial cover non-crop | 0 | 128.2 | 128.2 | (100.0) | 302.1 | 302.1 | (100.0) | 0.0 | 0.0 | (–100.0) |
| Alfalfa | 89.7 | 0.0 | –89.7 | (–100.0) | 400.9 | 311.1 | (346.5) | 0.0 | –89.7 | (–100.0) |
| Pasture | 118.6 | 0.0 | –118.6 | (–100.0) | 826.7 | 708.0 | (596.9) | 0.9 | –118.6 | (–100.0) |
| Pond/lake | 3.7 | 3.6 | –0.1 | (–4.3) | 6.1 | 2.3 | (61.9) | 3.7 | –99 | (–0.3) |
| Fencerow | 67.4 | 0.2 | –67.2 | (–99.7) | 58.5 | –8.8 | (–13.2) | 0.0 | –67.3 | (–99.9) |
| Riparian areas | 25.1 | 46.7 | 21.5 | (85.8) | 72.9 | 47.7 | (189.6) | 263.2 | 238.0 | (945.1) |
| Intercropping | 0.0 | 0.0 | 0.0 | (0.0) | 0.0 | 0.0 | (0.0) | 1931.7 | 1931.7 | (100.0) |
| Biodiversity garden | 0.0 | 0.0 | 0.0 | (0.0) | 84.1 | 84.1 | (0.0) | 91.5 | 91.5 | (100.0) |
| Organic crops | 0.0 | 0.0 | 0.0 | (0.0) | 0.0 | 0.0 | (0.0) | 153.8 | 153.8 | (100.0) |
| Permanent bioreserve | 0.0 | 0.0 | 0.0 | (0.0) | 0.0 | 0.0 | (0.0) | 91.4 | 91.4 | (100.0) |
| Wetland | 0.0 | 0.0 | 0.0 | (0.0) | 0.0 | 0.0 | (0.0) | 197.9 | 197.9 | (100.0) |
| Other | 343.1 | 285.3 | –57.8 | (–16.8) | 352.4 | 9.3 | (2.6) | 400.9 | 57.8 | (14.4) |
| Total land | 5,138 | 5,138 | | | 5,138 | | | 5,138 | | |

*Note*: This study was conducted using metric measurements. Each hectare (ha) equals 2.47 acres.

**Table 7-2.** Land Cover Change in Buck Creek

| Land cover | Baseline Area, (ha) | Scenario 1 Area, (ha) | Change, (ha) | Change, % | Scenario 2 Area, (ha) | Change, (ha) | Change, % | Scenario 3 Area, (ha) | Change, (ha) | Change, % |
|---|---|---|---|---|---|---|---|---|---|---|
| **Woodland closed** | 406.1 | 210.7 | −195.3 | (−48.1) | 60.6 | −345.4 | (−85.1) | 582.4 | 176.2 | (43.4) |
| **Woodland open** | 240.4 | 98.0 | −142.4 | (−59.2) | 171.4 | −69.0 | (−28.7) | 235.2 | −5.1 | (−2.2) |
| **Savannah** | 126.3 | 59.4 | −66.8 | (−52.9) | 123.1 | −3.1 | (−2.5) | 155.6 | 29.3 | (23.2) |
| **Corn/soybeans** | 3823.3 | 5277.1 | 1453.8 | (38.0) | 1063.1 | −2760.1 | (−72.2) | 88.2 | −3735.0 | (−97.7) |
| **Perennial cover non-crop** | 1392.2 | 1919.2 | 526.9 | (37.8) | 371.6 | −1020.6 | (−73.3) | 0.0 | −1392.2 | (−100.0) |
| **Alfalfa** | 353.8 | 590.2 | 236.3 | (66.8) | 3640.9 | 3287.0 | (928.8) | 0.0 | −353.8 | (−100.0) |
| **Pasture** | 1217.9 | 0.0 | −1217.9 | (−100.0) | 1689.4 | 471.5 | (38.7) | 0.0 | −1217.9 | (−100.0) |
| **Pond/lake** | 33.5 | 31.7 | −1.7 | (−5.4) | 38.8 | 5.3 | (15.8) | 41.6 | 8.0 | (24.1) |
| **Fencerow** | 123.4 | 0.0 | −123.4 | (−100.0) | 366.4 | 243.0 | (196.8) | 0.0 | −123.4 | (−100.0) |
| **Riparian area** | 51.6 | 95.1 | 43.5 | (84.4) | 384.4 | 332.8 | (644.8) | 1113.8 | 1062.2 | (2057.6) |
| **Intercropping** | 0.0 | 0.0 | 0 | (0.0) | 0.0 | 0.0 | (0.0) | 5432.1 | 5432.1 | (100.0) |
| **Biodiversity garden** | 0.0 | 0.0 | 0 | (0.0) | 80.0 | 80.0 | (100.0) | 117.2 | 117.2 | (100.0) |
| **Organic crops** | 0.0 | 0.0 | 0 | (0.0) | 0.0 | 0.0 | (0.0) | 28.0 | 28.0 | (100.0) |
| **Permanent bioreserve** | 0.0 | 0.0 | 0 | (0.0) | 0.0 | 0.0 | (0.0) | 233.4 | 233.4 | (100.0) |
| **Wetland** | 0.0 | 0 | 0.0 | (0.0) | 0.0 | 0.0 | (0.0) | 0.0 | 0.0 | (100.0) |
| **Other** | 1051 | 538.6 | −512.4 | (−48.8) | 830.3 | −220.7 | (−21.0) | 792.5 | −285.5 | (−24.5) |
| **Total land** | 8,820 | 8,820 | | | 8,820 | | | 8,820 | | |

*Note:* This study was conducted using metric measurements. Each hectare (ha) equals 2.47 acres.

been monitored for water quality each month since 1990. Nitrate concentrations during that period often exceeded 10 mg/L, and annual watershed nitrate export ranged from 4 to 59 lb/acre (4 to 66 kg/ha; Cambardella et al. 1999; Hatfield et al. 1999). These losses represented 4 to 115 percent of the nitrogen fertilizer applied each year (Jaynes et al. 1999).

Data were collected in the Buck Creek watershed between March 1998 and June 1999. A recording pressure transducer (Solinst 3400, EQUIPCO, Concord, California) was installed in July 1998, and a rating curve was developed to generate hourly discharge data. Rainfall data were collected with a standard tipping bucket gauge and an Onset® data logger. The daily Buck Creek hydrograph was extended to a synthetic eight-year data set, using a transform function based on a linear regression with data collected by the U.S. Geological Survey (USGS) near Hartwick, Iowa (USGS Station Number 05452200; Shoup 1999). Water samples were collected from late winter to early spring in 1998 and 1999, with multiple samples collected during periods of high discharge. Grab samples were collected monthly, and an ISCO™ 6700 sampler was used to collect close interval samples during six precipitation and snowmelt events, representing periods of rapidly fluctuating discharge. Samples were transported in iced coolers to the analytical laboratory and analyzed for total suspended solids (TSS), pH, specific conductance, nitrate + nitrite nitrogen as N, and total phosphate as P. Table 7-3 summarizes nitrate-N and TSS data for the 116 samples collected during this time period. For the 116 samples, nitrate concentrations in Buck Creek never exceeded 10 mg/L even in samples collected during the spring and summer, when nitrate concentrations tend to be greatest. In Walnut Creek, though, nitrate concentrations exceeded 10 mg/L about 30 percent of the time between 1992 and 1995. Data from the analyses for TSS are also presented in Table 7-3. The median TSS value was approximately 1 g/L, with a maximum of 26 g/L. These highly elevated values indicate that erosion is a significant problem in the Buck Creek watershed. Because sample collection targeted storm events, these values likely represent maximum values for TSS in this watershed. During high flows, nitrate concentrations decreased because of dilution and TSS concentrations increased because of the increased erosive potential of the higher discharges. These patterns are consistent with those observed in similar studies.

## Using the SWAT Model

The SWAT model is a continuous, spatially explicit simulation model designed to quantify the effects of land use and management change on water quality in agricultural basins (Arnold et al. 1995). The version of the model we used incorporated a geographic information system (GIS) interface and used the Hydrologic Response Unit configuration (Manguerra and Engel 1998; Di Luzio et al. 2000). The GIS interface simplifies the process of watershed discretization and parameter assignment by using spatially explicit data sets that represent elevation surface, soils, land use, and management.

## Evaluating Our Results

The scenarios evaluated in this project were designed with different emphases on policy and agricultural practices. The model was used to quantify the improvement that might be expected with the implementation of each scenario in each of the two study watersheds. Each scenario was simulated for both watersheds on a daily basis for an eight-year period, and summarized by year for seven years from 1992 through 1998, inclusive. We present the results from these simulations as percentage changes in me-

**Table 7-3.** Summary of TSS and Nitrate-N Data for Buck Creek and Walnut Creek Outlet Stations

| | Mean | Standard Deviation | Median | Maximum | n | % > 10 mg/L |
|---|---|---|---|---|---|---|
| **Buck Creek** | | | | | | |
| Nitrate-N, mg/L | 6.02 | 1.85 | 6.45 | 9.6 | 116 | 0 |
| TSS, mg/L | 2208 | 3939 | 1012 | 27200 | 116 | n/a |
| **Walnut Creek** | | | | | | |
| A. Nitrate-N, mg/L | 8.0 | 3.6 | 8.3 | 20.9 | 966 | 33 |
| B. Nitrate-N, mg/L | 9.0 | 3.4 | 9.3 | 20.9 | 628 | 45 |

*Notes*: Rows A and B for Walnut Creek contrast annual data with spring–summer data, with Row A representing all MSEA data collected between 1992 and 1995. Row B represents all MSEA data taken during March, April, May, June, and July between 1992 and 1995. The abbreviation n/a indicates that these data were not available for this watershed.
*Source*: Walnut Creek data provided by J. Hatfield.

dian yearly N and TSS loading for the simulated period. We used the median values in an effort to quantify water quality response to changes in land use without emphasizing extreme events.

### Erosion

Modeled results indicated that upland erosion and TSS concentrations would decrease for all scenarios in both watersheds relative to baseline conditions (Figures 7-1 and 7-2). TSS concentrations forecasted for Scenarios 2 and 3, however, were substantially lower than those for Scenario 1.

The major difference between the baseline landscape and Scenario 1 is the basin-wide implementation of conservation tillage. The median TSS loading in each of the two streams decreases by about 15 percent, whereas the area in corn and soybean production increases from 80 to 86 percent of the total watershed area in Walnut Creek and from 43 to 62 percent in Buck Creek. These results suggest that moderate reductions in soil loss could be achieved by widely implementing conservation tillage. Simulated decreases in erosion from Scenarios 2 and 3 were significantly greater, ranging from 35 to 60 percent reductions in the median sediment yield. We attribute this improvement to the implementation of a more complete set of management practices, including increased perennial cover combined with decreased production of corn and soybeans. The TSS loadings modeled under Scenario 2 exhibited the greatest decline from baseline values, with similar declines shown in Scenario 3. The combinations of alternative practices embodied in Scenarios 2 and 3 (for example, 30-60 m riparian buffers, strip intercropping, carefully managed rotational pasture, and large areas set aside as habitat reserves or in perennial cover) appear to be effective in reducing the erosion and sediment export compared to corn and soybean row crops. Bioreserve set-asides and alternative crops under Scenario 3 are not simulated to decrease erosion more than maintenance of perennial cover on erodible land under Scenario 2, either in the form of alfalfa or hay fields or as carefully managed rotational pasture.

The pattern of erosion reduction is similar in both watersheds, but the magnitude of the decrease is simulated to be larger in Buck Creek. Buck Creek is a well-developed

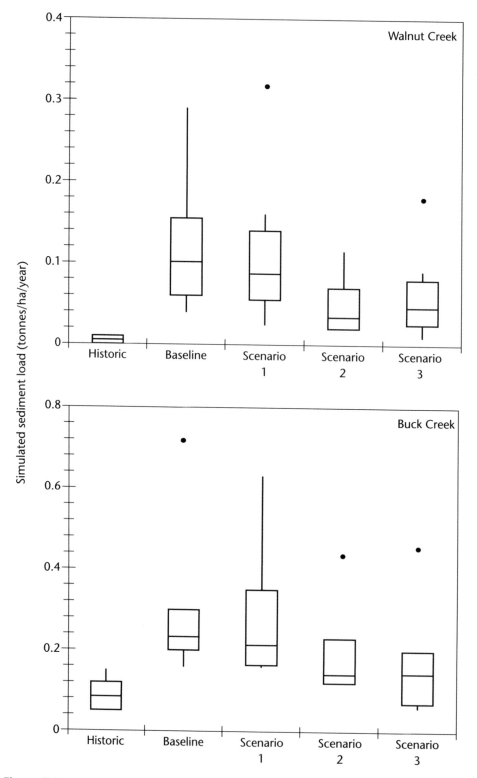

**Figure 7-1.** Boxplot Representing Sediment Loading in Walnut and Buck Creek Watersheds

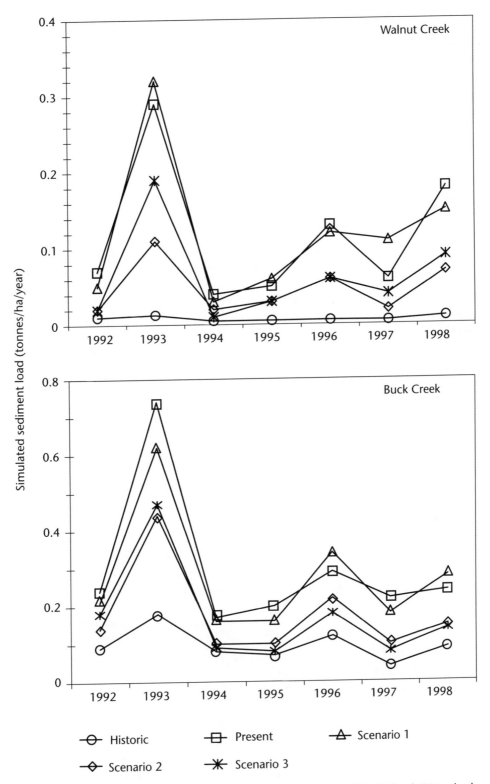

**Figure 7-2.** Time Series Plots of Sediment Loading in Walnut and Buck Creek Watersheds

*Note*: The time series help explain the variation noted in Figure 7-1, particularly the high values which occurred in 1993, a year of widespread flooding throughout the Midwest.

stream system with much greater relief than Walnut Creek. Overland flow moves over steeper terrain in Buck Creek, which increases its potential to erode. Additionally, upland flow paths tend to be shorter in Buck Creek because of the dendritic nature of the channel. These shorter flow paths result in higher TSS values at the base of the watershed, as sediment is less likely to be redeposited before it reaches the stream.

## Nitrate

The results for the nitrate simulations differ from those for TSS (Figures 7-3 and 7-4). Under Scenario 1, nitrate concentrations are modeled to increase in both watersheds. In this scenario, the area in monoculture production of corn and soybeans increases in both watersheds, but the management assumption is that current application rates of nitrogen fertilizer do not change. Fertilizer applications are expected to be targeted to locations where they will produce the greatest yield increases, but average amounts of nitrogen applied per acre or hectare of cropland are expected to remain about the same. The greater mass of applied nitrogen results in increased nitrate runoff for Scenario 1.

The decreased nitrate loading under Scenarios 2 and 3 mimics that of TSS, with significant decreases forecast in both cases. Reductions in the median nitrate load range from 57 to 70 percent. In Scenario 2, much of the Walnut Watershed is still in corn and soybean production (Table 7-1) but not in continuous corn and soybean rotation. Instead, with grazing livestock integrated into farming enterprises, a corn–soybean–alfalfa rotation is assumed. In our SWAT model, we assumed this rotation includes two years of unfertilized alfalfa, and application of nitrate to corn reduced by a modest 10 percent (to 120 kg/ha/yr of rotation as anhydrous ammonia). Additionally, a significant area of both watersheds (988 acres [400 ha] in Walnut Creek and 8,995 acres [3,640 ha] in Buck Creek) is converted to alfalfa production, and pasture land increases substantially in Scenario 2. During the simulation period, much less commercial nitrate is applied to these land covers. As a result, significantly less nitrate is simulated as being exported from the watershed in the stream system. Results from Scenario 3 suggest significant decreases in nitrate export compared to the baseline landscape, although, as with TSS, the largest improvements occur under Scenario 2. Under Scenario 3, the area in corn and soybean production in both watersheds is greater than under Scenario 2, but reduced compared with Scenario 1, replaced primarily with bioreserves, riparian areas, and perennial grasses in strip intercropping. As a result, nitrate runoff is consistently higher in Scenario 3 than in Scenario 2, but much less than under Scenario 1.

In Buck Creek, the Scenario 3 design rules result in an extreme decrease in corn and soybean monocultures (from 9,447 acres [3,823 ha] in the baseline landscape to 217 acres [88 ha]). This decrease is due in large part to conversions from corn and soybean monocultures to perennial strip intercropping, including about two-thirds of its area in corn and bean strips, as well as to the implementation of a 197-ft (60-m)-wide riparian buffer along perennial streams. These changes convert a major portion of the Buck Creek watershed from conventional row crops. The dendritic nature of the stream system in Buck Creek gives this 60-m buffer an area of 2,752 acres (1,114 ha) or approximately 13 percent of the basin area. In Walnut Creek, because the stream network is less complex, the riparian buffer system in Scenario 3 is approximately 649 acres (263 ha), which is about 5 percent of the watershed area. Additionally, in Walnut Creek the area in corn and soybean production (including perennial strip intercropping) in Scenario 3 is higher than that in Buck Creek. Walnut Creek has 9,034 acres (3,656 ha) or 70 percent

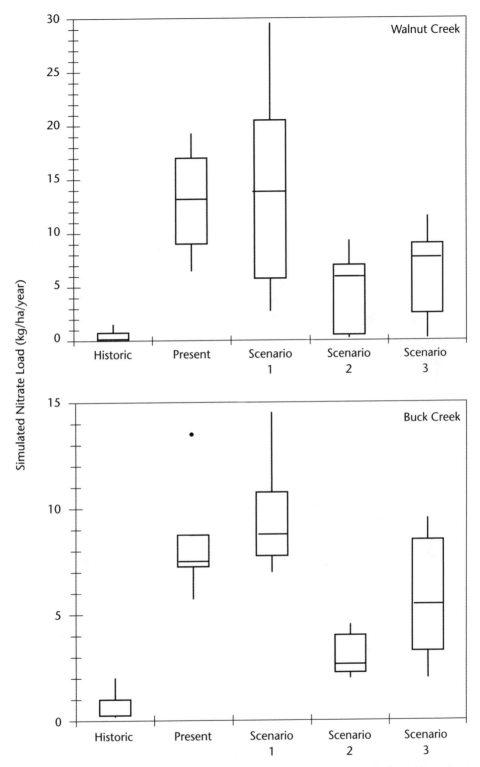

**Figure 7-3.** Boxplots Representing Nitrate Loading in Walnut and Buck Creek Watersheds

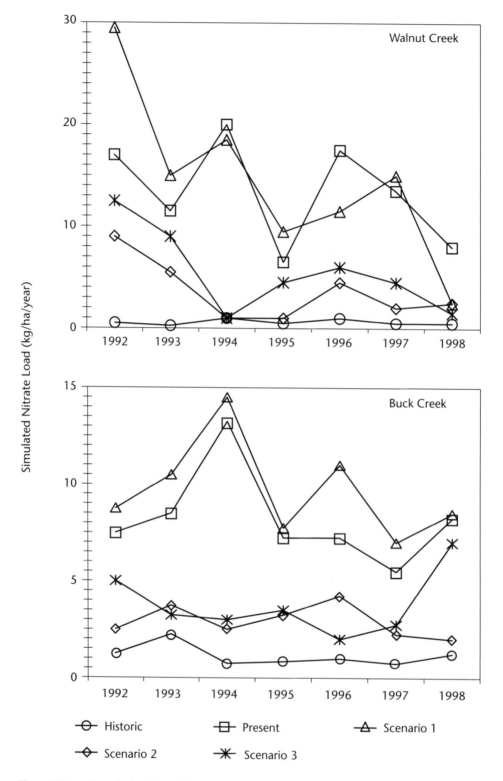

**Figure 7-4.** Time Series Plots of Nitrate Loading in Walnut and Buck Creek Watersheds for the Study Period

*Note*: The time series help explain the variation noted in Figure 7-3.

of its area in enterprises that produce corn and soybeans, whereas 13,640 acres (5,520 ha; 62 percent) of Buck Creek is under corn and soybean production (intercropped with perennials) in Scenario 3. As a result, significantly less nitrate is applied to Buck Creek. Even given its larger area—it is 59 percent larger than Walnut Creek—model results and calibration data indicate that loadings of nitrate for Buck Creek are generally smaller than for Walnut Creek.

### *Comparison of Historical and Current Landscape*

Simulated water quality under historical conditions differed significantly from current conditions. The simulations using predevelopment land cover yielded values representing reductions of 90 and 96  percent for TSS and 96 and 87 percent for nitrate in Walnut Creek and Buck Creek, respectively. These large differences result from three related factors.

The first is reconstructed historical land cover. Native prairie ecosystems were not tilled and the relatively dense perennial cover reduced erosion potential and TSS concentrations. In addition, the historical reconstruction for Walnut Creek had 5,940 acres (2,404 ha) of wetlands, which represents 46 percent of the total watershed area and greatly alters the watershed's hydrology. Estimated historical wetland area in Buck Creek is considerably less than that in Walnut Creek, mostly because of the greater topographic relief. As a result, estimates of historical nitrate and sediment loading in Buck Creek are higher than those for Walnut Creek.

The second factor involved in the significantly improved water quality under historical land cover is that chemical fertilizers were not added to the systems, and consequently, nutrient export was considerably less. Finally, these watersheds historically stored more water in upland wetlands, significantly reducing peak flows and development of stream channels. This reduced the potential for erosion and sediment delivery to streams.

## Water Quality Improvements

Our study results lend support to an emerging view that achieving substantial reductions (of approximately 30 to 75 percent) in nutrient and sediment export in areas such as the Corn Belt will require major modifications of agricultural practices as well as wetland restoration along higher-order streams and in lowlands (Chapter 15). In our study, simulated reductions in nitrogen export of >50 percent resulted from changes in land use and management in which nitrogen applications decreased by 10 to 33 percent, along with implementation of other BMPs in Scenarios 2 and 3.

Scenario 1, in which current BMPs were employed without substantially reducing the total amount of nitrogen applied, showed increased levels of nitrate export. These results indicate that continued use of baseline levels of nitrogen to fertilize crops will continue to result in nitrogen export to surface and groundwater that is greater than or equal to current values and substantially greater than historical values. Chapter 5 models the same future scenarios for Walnut Creek watershed using the Interactive Environmental Policy Integrated Climate (i_EPIC) model, and those results also showed an increase in nitrate runoff in Scenario 1.

Conversion to no-till cultivation and residue management across large areas, as envisioned in Scenario 1, may not by itself effectively reduce sediment concentrations. Only modest reductions in TSS were realized with exclusive no-till cultivation, whereas reductions of up to 65 percent occurred when other changes (including decreased row crop production and wider riparian areas) were also incorporated under other scenarios.

## Implications for Agricultural Policy

Our research adds to a body of literature (Wolf 1995; Schilling and Thompson 2000; Becher et al. 2000; Mitsch et al. 2001) indicating that water quality improvement on the order of 30 to 50 percent loading reductions will require land managers to develop fully integrated watershed plans including innovative land use and management practices. Because such practices require some cropland to be taken out of production, agricultural policy and public support for such practices must be a part of these plans, as assumed in our scenarios.

The two watersheds in Iowa modeled in this study are representative of the Des Moines Lobe and the Southern Iowa Drift Plain physiographic provinces. We encourage further testing of these ideas, both with modeling and with watershed-level experiments (see, for example, Likens et al. 1977; Boody et al. 2005), on a wider selection of watersheds in other physiographic and agricultural settings. Such research is needed to understand the extent, costs, and benefits of various practices aimed at improving water quality at multiple scales, in agricultural regions, and in metropolitan areas downstream.

## Acknowledgments

We acknowledge support from the U.S. Environmental Protection Agency (U.S. EPA) STAR grants (Water and Watersheds) program under grant R825335-01. We thank Michael Shoup and Dr. Jerald Schnoor of the University of Iowa for assistance in the Buck Creek sampling, and Dr. Jerry Hatfield of the USDA Agricultural Research Service (ARS) Tilth Laboratory for supplying the data for Walnut Creek. We thank the American Water Resources Association for permission to reprint material from Vaché et al. (2002). Although this work was funded by the U.S. EPA, the conclusions and opinions presented here are solely those of the authors and are not necessarily the views of the agency.

# Chapter 8

# Plant Diversity

Mary V. Santelmann, Jean C. Sifneos, Denis White,
and Kathryn Freemark Lindsay

Of the world's ecosystems, grasslands and plains have undergone some of the greatest transformations as a result of agriculture. In Iowa, approximately 99 percent of the lands historically in prairie, 99.9 percent of the wetlands, and more than 75 percent of the forest have been converted to cropland or pasture (Bishop 1981; Farrar 1981; Thompson 1992). Much of the remaining natural vegetation has been altered, fragmented, and isolated from other natural remnants by a matrix of cropland (Roosa 1981).

Changes in biogeochemical cycles and ecosystem processes have also contributed to the decline of native plant diversity through their influence on landscape evolution and vegetation dynamics. The hydrologic response to precipitation in the region is now relatively rapid, and flow occurs along a few linear flow paths, leading to channel incision. Nutrient flux through the system is several times greater than that of the past, and soils are being "mined" of both organic carbon and nutrients (Mitsch et al. 2001; Vaché et al. 2002). Wetlands, floodplains, riparian zones, and roadsides, which often contain the last remnants of natural vegetation in these landscapes, are vulnerable to invasion by species such as reed canary grass, *Phalaris arundinacea* (L.) and purple loosestrife, *Lythrum salicaria*. The disturbance regimes characteristic of the region have been altered, potentially disrupting the life cycles of species adapted to historical disturbances such as fire or flood. Some native species require such disturbances to provide the conditions for germination, seedling establishment, or modification of competitive interactions that allow species to coexist. Introduced species also play a role in the loss of native species diversity. Approximately one-fourth of the nearly 1,300 species in the current flora of Iowa are introduced weeds (Withers et al. 2003).

The combination of conversion of native vegetation to cropland or pasture; introduction of non-native, invasive species; and changes in ecosystem processes have taken their toll on Iowa's native plant diversity. Of more than a thousand vascular plant species once found in the forests, prairies, and wetlands of Iowa, 45 species have been extirpated, and an additional 221 species have been listed as endangered, threatened, or of special concern (Roosa et al. 1989).

Here, we present the results of an approach for estimating the potential changes to native plant diversity that could result from changes in land use and management in Iowa embodied in the alternative future scenarios designed by Nassauer et al. (Chapter

4). We applied methods developed by White et al. (1997) to compare the plant diversity of alternative future landscapes that could result from three different scenarios of public support and policy. In these scenarios, priority is given, respectively, to increasing agricultural commodity production (Scenario 1), improving water quality and reducing downstream flooding (Scenario 2), and enhancing native biodiversity (Scenario 3) as described in Chapter 4, for Buck Creek and Walnut Creek watersheds in Iowa (as seen in Color Figures 38–43).

## Conducting the Study

Few methods exist for predicting and/or comparing potential plant diversity at the spatial scale of our watersheds. The methods that do exist were developed either to estimate current diversity based on sampling data from a "snapshot in time" for regions being managed primarily as natural areas (Stohlgren et al. 1997) or to estimate continental-scale diversity (Withers et al. 2003), rather than to compare relative floristic diversity of alternative landscapes. Our goals were (1) to quantify the potential impact of changes embodied in the alternative future scenarios on potential diversity of the flora in the two study watersheds, and (2) to suggest possible trends in the potential floristic diversity of landscapes of central Iowa under these alternative future scenarios.

We used a statistical measure of change in land cover weighted by an index of the frequency of occurrence of each plant species of each land cover type (from 0 = absent to 4 = common) to compare the potential diversity of the landscape futures for each scenario for all vascular plant species in central Iowa (after White et al. 1997 and Chapter 13). The formula used for calculating the change statistic, $HC_j$, for a specific group of species for each of the landscape futures or for the past landscape, was

$$HC_j = median \left( \sum_i^{species_j} \left( hab_{i,j} - hab_{i,present} \right) / hab_{i,present} * 100 \right)$$

In this formula, $j$, $hab_{i,j}$ is the frequency-weighted area of "habitat" for species $i$ in the future or past landscape $j$, and $hab_{i,present}$ is the frequency-weighted area of habitat for species $i$ in the baseline landscape. Positive values of the change statistic mean that more area of vegetation in which the species is commonly found occur in the watershed in the future or past landscape than in the baseline landscape, and negative values mean the reverse.

Data needed for this method were total area in each land cover type in each alternative landscape (obtained from the geographic information systems [GIS] database for the baseline landscape and each future scenario), and a species–land cover association matrix for the plants of central Iowa (described in the next section). This method is based on the premise that the impact on a species increases as the area of land cover in which that species occurs is converted or degraded.

### Developing the Database on Plant Distribution and Ecological Associations

We developed a species–land cover association matrix for all vascular plant species of central Iowa by downloading the list of Iowa species from the U.S. Department of Agriculture (USDA) Natural Resource Conservation Service (NRCS) PLANTS Database in 1997. The state plant list data are currently available on the USDA NRCS Web site (2007). Next, we reduced the list to 1,247 species by limiting inclusion to only those species occurring in the region of the study watersheds: Boone, Dallas, Hardin, Hamilton, Grundy, Jasper, Marshall, Polk, Poweshiek, Story, Tama, and Webster counties of central Iowa (Eilers and Roosa 1994). Distribution and assignment to land cover types were then cross-checked using data from the primary literature on central Iowa prairie, forest, and wetland vegetation (Kucera 1952; Russell 1956; Freckmann 1966; Niemann and Landers 1974; Lammers and van der Valk 1977a, 1977b; Johnson-Groh 1985; Niemann 1986; Thompson 1992; Jacobsen et al. 1992; Van der Linden and Farrar 1993; Galatowitsch and van der Valk 1994, 1996; USDA NRCS 1997). We then ranked each species for its abundance in the land cover classes present in these watersheds.

Initially, we coded species to the seven land cover types most commonly specified in the literature: forest, prairie, wetland, farmstead, roadside and fencerow, pasture, and cropland, and assigned a frequency of occurrence ranking (0 = absent, 1 = rare [<25 percent], 2 = infrequent [25–50 percent], 3 = frequent [50–75 percent], and 4 = common [>75 percent]). In addition, each species was coded (present =1, absent = 0) to a range along a moisture gradient: xeric, mesic, moist, and wetland. Regional experts then reviewed and edited the matrix. The expert-reviewed database can be downloaded from the Web at http://www.epa.gov/wed/pages/staff/white/getbiod.htm (U.S. EPA 2006). The final matrix used to describe plant associations with the land cover and vegetation types found in the alternative futures was expanded from the initial set to the classes shown in Table 8-1. We distinguished wet prairie species from dry prairie species by sorting the entire list of prairie species according to the moisture designations. Prairie species coded only to relatively xeric and mesic soils were assigned to upland prairie, keeping their original abundance code. Prairie species coded only to moist and wet soils were assigned to wet prairie, again keeping their original abundance code. Species coded to both wet prairie and wetlands were assigned to wet prairie with the prairie score and to wetlands with the wetlands score. Finally, species coded as common (4) or frequent (3) in prairie and coded to moist and wet soils as well as mesic and dry soils were assigned to upland prairie as frequent and to wet prairie as infrequent (2) or rare (1). All herbaceous species found in upland prairie were coded to savanna with the same scores, as were forb and tree species specifically noted by Eilers and Roosa (1994) as characteristic of open, upland forest.

## Monte Carlo Methods for Estimating Effects of Uncertainty

To investigate the effects of uncertainty in the species frequency of occurrence scores, we conducted a Monte Carlo simulation study in which frequencies of occurrence codes were altered under an assumed error model. Here, we assumed the scores to vary about their original values in accordance with a normal distribution with a standard deviation of 1. Next, we computed the variability in the results, and added the errors generated according to the error model to the original scores. We repeated the error generation process 1,000 times and calculated the mean and standard deviation to provide a measure of uncertainty in the results caused by possible errors in the plant abundance scores.

## Species–Area Relationships

Two published studies (Freckmann 1966; Galatowitsch and van der Valk 1996) were sources of data suitable for constructing species-accumulation curves for prairie and wetland sites in the Corn Belt region. These authors supply floristic inventories of five native prairie and ten wetland remnants between 3 and 22 acres (1.2 and 9 ha) in

**Table 8-1.** Land Cover Classes and Area Corresponding to Each Vegetation Type in Baseline and Past Landscapes and in Each Alternative Future Scenario for the Two Watersheds

| Vegetation Type | Land Cover Classes[a] | Past[b] | | Baseline[b] | | Scenario 1[b] | | Scenario 2[b] | | Scenario 3[b] | |
|---|---|---|---|---|---|---|---|---|---|---|---|
| | | Buck | Walnut | Buck | Walnut | Buck | Walnut | Buck | Walnut | Buck | Walnut |
| Forest | for.50 for.fpug for.upug for.upg | 4104 | 156 | 669 | 193 | 319 | 133 | 1098 | 290 | 2297 | 602 |
| Savanna | for.sug for.sg | 35 | 14 | 126 | 41 | 59 | 24 | 123 | 40 | 156 | 35 |
| Cropland | rc.cp rc.ct rc.fall rc.sg rc.ns[c] rc.os rc.o | 0 | 0 | 3823 | 4191 | 7196 | 4639 | 1063 | 2882 | 5548 | 3811 |
| Pasture, grass, and alfalfa/hay | shrub.past shrub.ung grass.crp[c] grass.hay | 0 | 0 | 2509 | 289 | 2509 | 39 | 5702 | 1561 | 0 | 35 |
| Farmstead | farmstead[c] | 0 | 0 | 122 | 179 | 53 | 117 | 83 | 132 | 120 | 148 |
| Roadside/ Fencerow | strip.h[c] strip.w[c] | 0 | 0 | 423 | 170 | 316 | 112 | 518 | 156 | 228 | 144 |
| Dry prairie | grass.pd | 4530 | 2558 | 0 | 0 | 0 | 0 | 0 | 0 | 233 | 90 |

**Table 8-1.** Land Cover Classes and Area Corresponding to Each Vegetation Type in Baseline and Past Landscapes and in Each Future Alternative Scenario for the Two Watersheds (continued)

| Vegetation Type | Land Cover Classes[a] | Past[b] | | Baseline[b] | | Scenario 1[b] | | Scenario 2[b] | | Scenario 3[b] | |
|---|---|---|---|---|---|---|---|---|---|---|---|
| | | Buck | Walnut | Buck | Walnut | Buck | Walnut | Buck | Walnut | Buck | Walnut |
| Wet prairie | grass.pw | 51 | 2206 | 0 | 0 | 0 | 0 | 0 | 0 | 0 | 171 |
| Wetland | wet.sp | 6 | 182 | 0 | 0 | 0 | 0 | 0 | 0 | 0 | 27 |
| Pond | wet.pond | 20 | 4 | 34 | 4 | 32 | 4 | 34 | 5 | 42 | 4 |
| | wet.st | | | | | | | | | | |
| | wet.eng | | | | | | | | | | |

[a] These land cover classes correspond to those used for modeling changes in vertebrate biodiversity in Chapter 13 (Table 13-1). They include row crop chisel plow (rc.cp), row crop conservation tillage (rc.ct), row crop native strip (rc.ns), small grains (rc.sg), fallow (rc.fall), row crop organic strip (rc.os), row crop organic (rc.o), farmstead, herbaceous strip (strip.h), woody strip (strip.w), Conservation Reserve Program or other perennial non-crop (grass.crp), alfalfa/hay (grass.hay), dry prairie (grass.pd), wet prairie (grass.pw), pasture (shrub.past), ungrazed shrubland (shrub.ung), ungrazed forest less than 50 years old (for.50), ungrazed riparian forest (for.fpug), ungrazed upland forest (for.upug), grazed upland forest (for.upg), ungrazed savanna (for.sug), grazed savanna (for.sg), semipermanent wetland (wet.sp), pond (wet.pond), stream (wet.st), and engineered wetland. (wet.eng)

[b] This study was conducted using metric measurements. Each hectare (ha) equals 2.47 acres.

[c] Planted with native species to enhance biodiversity in Scenario 3

**Table 8-2.** Median Percent Change and Standard Deviation in Potential Plant Diversity

| Landscape | Scenario 1 | | Scenario 2 | | Scenario 3 | | Past | |
|---|---|---|---|---|---|---|---|---|
| | % change | Std. dev. | % change | Std. dev. | % change | Std. dev. | % change | Std. dev. |
| **Buck Creek native species** | -52 | 0 | 49 | 1.2 | 208 | 0.2 | 514 | 0 |
| **Walnut Creek native species** | -31.9 | 0 | 42 | 1.3 | 212 | 0.2 | 262 | 22.4 |
| **Buck Creek introduced species** | -27 | 1.5 | 10 | 1.6 | -28 | 3.2 | -100 | 0 |
| **Walnut Creek introduced species** | -31 | 0 | -5 | 1.9 | -16 | 0.6 | -100 | 0 |

*Notes:* These results for native and introduced plant species are from a land cover-based model for alternative future scenarios in Walnut Creek and Buck Creek. In both watersheds, maximum potential native species richness ($S_{max}$) is 755 species for the land cover and associated vegetation that occurs in the baseline landscape and Scenarios 1 and 2, whereas $S_{max}$ = 932 for Scenario 3 and $S_{max}$ = 966 for the past.

area, a size typical for the region. We used these published data to conduct an analysis of species-area patterns for two of the most common native vegetation formations in central Iowa, upland prairies and prairie pothole wetlands.

The species accumulation curves we constructed from the published data serve as a first estimate of the extent to which individual sites might be expected to capture subsets of the species occurring in that vegetation type. They also help estimate the proportion of the species remaining in the set that are captured as additional sites of that specific vegetation type are added. The data are plotted first as average individual species richness for a single site, then as the average cumulative species richness for all combinations of two sites, then three sites, and so on up to the total of five prairie and ten wetland sites. The program PC-ORD Version 4.0 (MJM Software, Gleneden Beach, Oregon) was used for calculating the mean number of species accumulated by adding additional sites, and median site area (6.1 acres [2.5 ha] for wetlands and 12.3 acres [5 ha] for prairies) was used in plotting curves.

## Evaluating Our Results

The habitat change statistic, indicating greater native plant diversity, was highest for the alternative futures resulting from Scenario 3, with improvements relative to the baseline landscape of more than 200 percent for both Buck and Walnut Creek watersheds (Table 8-2). Scenario 2 ranked second, with 42 to 49 percent improvement. For Scenario 1, the habitat change statistic decreased by 32 to 52 percent.

### Potential Species Richness

Of the 971 native vascular plant species in our database for central Iowa, 34 have been extirpated and 5 were unlikely to be found in our study watersheds. Of the remaining 932 species, 177 are associated primarily with prairies, and were not expected to be found in the baseline landscape nor in Scenarios 1 or 2. Some of these species could potentially be restored in the bioreserves of Scenario 3, although restorations tend to be less diverse than native remnants (Galatowitsch and van der Valk 1996; Martin et al. 2005). We would expect that surrounding land use and site management would influence the eventual diversity of restorations (Thompson 1992; Galatowitsch and van der Valk 1994).

Forest, prairie, and wetland habitats have the highest species richness. In the species database, 386, 352, 335, and 187 native species were coded to forest, dry prairie, wet prairie, and wetlands, respectively. Only 72 to 164 native species were coded to cropland, pasture, or farmsteads. In addition, 89 to 91 percent of all the species coded to forest, prairie, or wetland were native to Iowa, compared to only 36 to 38 percent for cropland and farmsteads and 51 to 54 percent for fencerow and pasture, respectively.

Both species richness and equitability or evenness of distribution are components of diversity. Although we cannot precisely quantify the equitability of plant species distributions in future landscapes, the plant database can help characterize this attribute for the vegetation types common in those landscapes. Forest, prairie, and wetland areas had the highest proportion of native species coded as common (occurring in >75 percent of sites) with 24, 14, and 12 percent of the native species considered common, respectively. In contrast, only 1 to 4 percent of the native species occurring in cropland, pasture, or farmsteads are considered common. Consequently, landscapes that include more forest, prairie, and wetland habitats are likely to see higher species richness and greater equitability of species distribution than those dominated by cropland. Some vegetation types, such as wet prairies and fens, tend to harbor a large proportion of rare species, and these may be critical sites for conservation of rare species and vegetation types. For example, Pearson and Loeschke (1992) found that of 225 species found in a survey of 200 Iowa fens, 91 of these species occurred in fewer than 5 locations.

### Species-Accumulation Curves

The occurrence of a small tract of prairie or a single wetland in a watershed does not mean that all the plant species associated with prairies or wetlands will be found there. Species-accumulation curves can help elucidate patterns of diversity and expectations for the minimum number and type of sites required to capture a given percentage of the possible native plant diversity in a region (Stohlgren et al. 1997). The species-accumulation curves shown in Figure 8-1(a, b) approach the total number of species coded as nonrare for those vegetation types—approximately 202 species for dry to mesic prairie, 184 species for wet prairie, and 95 species for wetlands. This is consistent with the expectation that rare species will occur only infrequently in remnants or restorations, and that conservation of rare species will require species-specific protection plans. The pattern of increasing species richness with increasing area shown in the species-accumulation curves, though, helps to illustrate an important principle—that conservation of species now considered common may well depend on restoration efforts that furnish multiple sites for native species to grow, reproduce, and evolve in response to changing environments (Rosenzweig 1995). Restorations like the bioreserves in Scenario 3 can also help protect and buffer remaining remnants of native vegetation with plantings of native species and other management practices to minimize problems of invasive non-native species.

### Species–Area Relationship Curves

Larger areas set aside as reserves often comprise a mosaic of different vegetation types. Figure 8-2 shows the relationship between vascular plant species richness and area for remnants of native vegetation and several Iowa parks and preserves, many of which include several vegetation types. In this figure, species richness is simply plotted against the area of the park or preserve, analogous to a plot of species richness among islands in an archipelago (cf. Rosensweig 1995). In this case, the islands are native vegetation remnants in a "sea" of cropland. These data were fit to the log-transformed version of the Arrhenius equation ($S = cA^z$), which is commonly used to describe species–area relationships. This equation takes the form $\log (S) = \log (c) + z \log (A)$, with the following values obtained for the constants:

$$z = 0.38 \text{ and } c = 50.05 \ (r^2 = 0.75).$$

The species-accumulation curves in Figure 8-1(a, b) help illustrate the relationship between area and alpha diversity (within-habitat diversity) of small, noncontiguous patches of a single vegetation type characteristic of this region. In contrast, the species richness of successive points plotted in Figure 8-2 is not cumulative. Figure 8-2 is useful for comparing the species richness of small sites (which may or may not include more than one vegetation type) to the species richness found in large, heterogeneous mosaics of several natural vegetation types, such as the bioreserves larger than 640 acres (259 ha) in Scenario 3. Their diversity approaches the total diversity of the re-

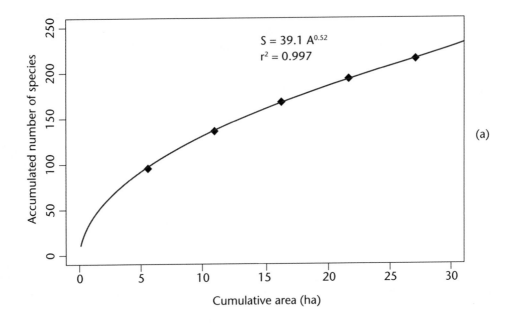

$S = 39.1\ A^{0.52}$
$r^2 = 0.997$

(a)

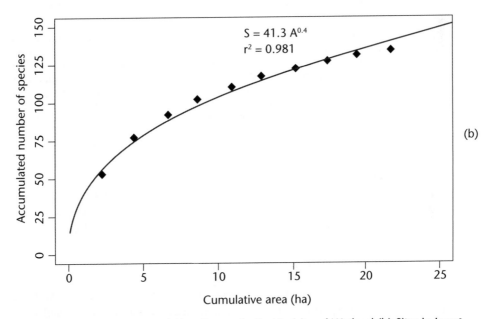

$S = 41.3\ A^{0.4}$
$r^2 = 0.981$

(b)

**Figure 8-1.** Species-Accumulation Curves for Prairie (a) and Wetland (b) Sites in Iowa[a]

[a] The prairie curve was generated from data in Freckman (1966) and the wetland curve from data provided by Galatowitsch (collected in studies published by Galatowitsch and van der Valk in 1996). These data were fit to the log-transformed version of the Arrhenius equation ($S = cA^z$) commonly used to describe species–area relationships, with $z = 0.52$ and $0.40$ and $c = 39.11$ and $41.28$ for prairie remnants and wetlands, respectively.

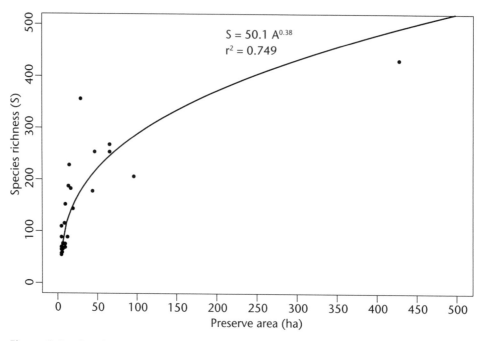

**Figure 8-2.**   Species–Area Relationships for Iowa Native Vegetation Remnants, Parks, and Preserves[a]

[a] Each point represents the species richness (S) of an individual site, park, or preserve of a particular area (A). Data were compiled from Freckmann 1966; Lammers and van der Valk 1977a, 1977b; Niemann and Landers 1974; Johnson-Groh 1985; Niemann 1986; Thompson 1992; Galatowitsch and van der Valk 1996, and Iowa DNR 2005. The two parks with the highest species richness are Ledges State Park (447) and Woodman Hollow (358), preserves that include forested areas as well as prairie openings and rock ledge habitats. These data were fit to the log-transformed version of the Arrhenius equation ($S = cA^z$) commonly used to describe species–area relationships: $\log (S) = \log (c) + z \log (A)$, with the following values obtained for the constants $z = 0.38$ and $c = 50.05$ ($r^2 = 0.75$).

gional flora, "gamma diversity" as described by Whittaker (1970), encompassing both the diversity inherent in a single patch of a given vegetation type and that of multiple vegetation types within a region. Even the largest of these preserves, however, contains fewer than 50 percent of the native plant species found in the regional flora.

## Implications for Agricultural Policy

As this investigation demonstrated for the Iowa study watersheds, current land use and management practices in existing agricultural landscapes of the Corn Belt have produced a landscape dominated by monocultures of non-native crop species, with massive losses of native plant biodiversity at the landscape level (Table 8-2).

Our results indicate that if public policy continues to emphasize agricultural commodity production as in Scenario 1, native plant diversity can be expected to continue its decline (for examples, see Color Figures 13 and 16). Introduced species would increase at the expense of native species in both watersheds (Table 8-1). If priorities include water quality improvement, giving rise to larger areas of more diverse vegetation, landscapes such as those in Scenario 2 may emerge. This scenario includes a more diverse vegetation physiognomy as well as row crops; (for examples, see Color Figures 14 and 17). The use of native species in perennial pasture mixtures could further increase their diversity. In Scenario 3, innovative native perennial strip intercropping systems increase diversity within fields of row crops. Under the same scenario, large bioreserves are likely to enhance the occurrence of native vegetation, and allow people the opportunity to appreciate that diversity (for examples, see Color Figures 15 and 18). This scenario, which outstrips all others in providing habitat for native plant species, is the scenario most likely to result in the kind of habitat mosaics required by specialist pollinators or wildlife species that require adjacent complementary habitat (see Chapters 9, 10, and 13).

At a regional scale, preservation in conjunction with restoration appears to be the best strategy for protecting and enhancing native plant biodiversity. Whether or not rare species will be found in central Iowa watersheds in 25 to 50 years will depend on efforts to preserve or restore ecosystems in specific locations where these rare species occur. The landscape-level diversity and richness of natural vegetation in these landscapes (which may provide wildlife habitat as well as water quality improvement), will reflect efforts to restore natural areas and incorporate native diversity into managed landscapes (Ehley 1992; Jacobsen et al. 1992).

## Acknowledgments

The authors acknowledge the Biota of North America Project (John Kartesz and staff) as a primary source for the USDA NRCS PLANTS database species list of plants found in Iowa, and *The Vascular Plants of Iowa* by Eilers and Roosa (1994) as the primary source for distribution and abundance information. We thank Mark Withers and his colleagues for supplying a list of references on the flora of Iowa. In addition, we are grateful to the following experts who reviewed the species lists, helped edit them for accuracy, and offered insights into the distribution and abundance of plant species in the land cover types found in the region: William Norris (forest and prairie species); Susan Galatowitsch (wetland species); Janet Thompson (forest species); Steven Barnhart (pasture, row crop, and fencerow/roadside species); and Douglas Buhner (row crop and fencerow/roadside species). We also thank reviewers M. Wilson and D. Olszyk for their comments and suggestions. The research described in this article has been subjected to the U.S. Environmental Protection Agency's peer and administrative review and approved for publication. Approval does not signify that the contents reflect the views of the agency, nor does mention of trade names or commercial products constitute endorsement or recommendation for use.

## Chapter 9

# Butterfly Responses

### Diane M. Debinski, Mary V. Santelmann, Denis White, Kathryn Freemark Lindsay, and Jean C. Sifneos

The impact of agriculture and of conservation programs on pollinators is important both economically and because they provide critical ecosystem services. Butterflies are a group of pollinators that are both diverse (in terms of the number of species) and relatively easily identified. They serve as excellent indicators of landscape change because they are responsive to vegetation change, often using very specific host plants in the larval stage. Butterflies have been significantly affected by habitat fragmentation in the Midwest (Schlict and Orwig 1998). Species that depend on native habitat, such as wetlands or prairies, are decreasing. In Iowa, two butterfly species are currently listed as endangered, five as threatened, and 25 as species of special concern (Iowa DNR 2005). Here, we consider how the three alternative scenarios envisioned by Nassauer, Corry, and Cruse (Chapter 4, this volume) in landscape futures for Buck Creek and Walnut Creek, Iowa, would affect the diversity of the butterfly community of the region.

### Conducting the Study

We examined the 117 species of butterflies currently recorded in central Iowa. After deriving 26 wildlife habitat classes from land cover classes, we generated a ranking of habitat suitability for each species in each potential habitat type. We reviewed the literature, especially Scott (1986), and used expert judgment to arrive at the rankings. Few species were unique to any single habitat type; many used a variety of habitat types. Some species, classified as habitat restricted, were modeled only to use prairies or wetlands.

The species–habitat associations were used to prepare a habitat map for each species in the past, the baseline, and for the landscape futures for each alternative scenario (Color Figures 30, 31, 34, 35, and 38–43). From these maps of habitat scores, we estimated the total amount of habitat for a species in a landscape as the sum of all the scores across the landscape. Next, we calculated the percentage change in habitat for each species relative to the baseline for the three future scenarios and for the past. Finally, we used the median of the percentage changes for different groups of species as a summary statistic, following the approach developed in White et al. (1997) and described in further detail in Chapter 13 of this book.

## Evaluating Our Results

Figure 9-1 illustrates the habitat change associated with each of the scenarios for 117 butterfly species modeled in both Buck and Walnut Creek drainages. As expected, the land use changes in Scenario 1 have the most negative effect on butterfly habitat relative to the baseline. Interestingly, the change between the past and the baseline or Scenario 1 is more extreme in the Walnut Creek watershed than in the Buck Creek watershed. Buck Creek, which has more hilly terrain, was less conducive to intensive agriculture than Walnut Creek and shows less dramatic loss in the baseline relative to the past. Compared with the baseline, Scenario 2 also has a larger habitat change score than Scenario 3 in Buck Creek. Because Buck Creek has more pasture and grassland and less area in row crops, Scenario 2 is more beneficial to several of the skipper species, which lay their eggs on grass. Total species richness differs among scenarios, but is the same for both watersheds with the exception of Scenario 3, in which the black dash (*Euphyes conspicua*) and two-spotted skipper (*E. bimacula*) are predicted to occur in Walnut Creek, with its large prairie-wetland reserve (Color Figure 43), but are not predicted to have preferred wetland habitat in Buck Creek.

Tables 9-1 and 9-2 summarize individual species that showed significant responses to the different scenarios. Of particular interest, both of the state's endangered species, the Dakota skipper (*Hesperia dacotae*) and the common ringlet (*Coenonympha tullia*), showed improvements under Scenario 3. Two species of special concern showed

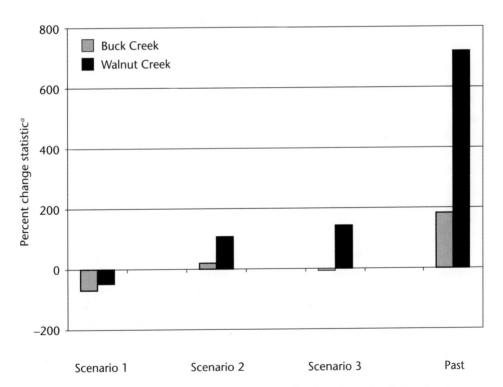

**Figure 9-1.** Changes in Potential Butterfly Diversity for the Alternative Future Landscapes

[a] The change statistic is a measure of the ratio of the future abundance relative to the baseline abundance of habitat for each species. A negative value signifies that less habitat is present under the future scenario than in the baseline landscape.

improvements under either Scenario 2 and Scenario 3 (*E. bimacula*) or Scenario 3 the arogos skipper (*Atrotone arogos*).

The differences between butterfly species that specialize in prairie, prairie/wetland, and wetland habitats are illustrated by differences among these three groups in Buck Creek between Scenarios 1 and 2. The prairie species are positively affected only in Scenario 3 (Color Figure 42), whereas the prairie/wetland species benefit from landscape changes under both scenarios. Wetland species are positively affected under Scenario 2 (Color Figure 40), which emphasizes buffers along streams and rivers and the creation of additional wetland area. These results make sense given the different priorities in land use under each of the scenarios.

In Walnut Creek, all three butterfly groups respond favorably to Scenario 3. The prairie/wetland species respond similarly to the wetland species and distinguish themselves from the prairie species in their differential responses to Scenario 2. Prairie species are not positively affected by Scenario 2 (Color Figure 41), but wetland or prairie/wetland species are positively affected. This response may be driven by the large prairie/wetland restoration in the western portion of the watershed and a riparian forest reserve along the creek.

The generalist species, fiery skipper (*Hylephillia phylleus*), which can be found on farmsteads, in hayfields, and in herbaceous buffer strips, had no habitat in the historic past in these watersheds. This species does best under Scenario 2 in both drainages, and it does better under Scenario 1 than Scenario 3 in Buck Creek. This species seems to be a good indicator of farm habitat.

## Influence of Land Use and Management Practices on Butterfly Diversity

Our results offer some insight as to how potential land use changes in future Corn Belt landscapes may affect pollinators. Adding additional habitat of filter strips, wetlands, and native perennial grasses positively affects butterfly species. Rare species are associated with both wetland and prairies, so both Scenarios 2 and 3 result in independently positive effects for Lepidoptera and increase the estimated species richness in these two watersheds. Percent change in habitat area, though, is just one metric to help in understanding future biodiversity patterns. Many additional variables (e.g., vegetation composition and structure, habitat management, edge effects, the shape of the habitat patch, and landscape context) will also influence butterfly distributions and abundance.

Vegetation composition is important because plant species provide hosts for larvae, nectar for adult butterflies, sites for laying eggs, and sites for overwintering. Shepherd and Debinski (2005a) found that the best overall vegetation predictor of butterfly richness and abundance was nectar availability, with some special considerations for specific groups. Reeder et al. (2005) found that butterfly abundance, including disturbance-tolerant species abundance, was best explained by forb cover and the number of flowering ramets. Abundance of habitat-restricted species, however, was best explained by the height and vertical density of the vegetation. These and other results (Collinge et al. 2003) indicate that habitat-restricted species may need tall grass for its habitat structure and specific microhabitat conditions.

Habitat management also influences butterfly communities. As an indicator of time since burning, litter has repeatedly proven to be an important factor influencing habitat-restricted species as well as overall species richness in both native and restored

**Table 9-1.** Influence of Landscape Change on Butterfly Species Richness for Buck Creek Watershed

| Species | Scenario 1 | Scenario 2 | Scenario 3 | Past | Habitat preference |
|---|---|---|---|---|---|
| Lycaena phlaeas | No habitat | No habitat | Habitat | Habitat | Prairie |
| Hesperia dacotae | No habitat | No habitat | Habitat | Habitat | Prairie |
| Atrytone arogos | No habitat | No habitat | Habitat | Habitat | Prairie, CRP |
| Coenonympha tullia | No habitat | Habitat | Habitat | Habitat | Prairie/wetland |
| Atrytone delaware | No habitat | Habitat | Habitat | Habitat | Prairie/wetland |
| Euphyes ruricola | No habitat | Habitat | Habitat | Habitat | Prairie/wetland |
| Euphyes conspicua | No habitat | Habitat | No habitat | Habitat | Wetland |
| Euphyes bimacula | No habitat | Habitat | No habitat | Habitat | Wetland |

**Table 9-2.** Influence of Landscape Change on Butterfly Species Richness for Walnut Creek Watershed

| Species | Scenario 1 | Scenario 2 | Scenario 3 | Past | Habitat preference |
|---|---|---|---|---|---|
| Lycaena phlaeas | No habitat | No habitat | Habitat | Habitat | Prairie |
| Hesperia dacotae | No habitat | No habitat | Habitat | Habitat | Prairie |
| Atrytone arogos | No habitat | No habitat | Habitat | Habitat | Prairie, CRP |
| Coenonympha tullia | No habitat | Habitat | Habitat | Habitat | Prairie/wetland |
| Atrytone delaware | No habitat | Habitat | Habitat | Habitat | Prairie/wetland |
| Euphyes ruricola | No habitat | Habitat | Habitat | Habitat | Prairie/wetland |
| Euphyes conspicua | No habitat | Habitat | Habitat | Habitat | Wetland |
| Euphyes bimacula | No habitat | Habitat | Habitat | Habitat | Wetland |

prairie (Shepherd and Debinski 2005a, 2005b; Skibbe 2005). Litter may also be an important contributor to microhabitat effects (e.g., sites in which to hide from predators, shade in the heat of the summer, and shelter for overwintering larvae).

Because prairies within the midwestern landscape can be viewed as refugia surrounded by a sea of agriculture, edge effects can be significant. Some species may be restricted in their movement because of their response to landscape features. For example, Ries and Debinski (2001) examined the behavior of two butterfly species at four different types of prairie edges (tree line, crop, field, and road) to determine how edges might act as a barrier to movement. The regal fritillary (*Speyeria idalia*) responded strongly to all edges, even those with low structural contrast such as a Conservation Reserve Program (CRP) field, and often avoided crossing an edge or returning to the prairie after crossing an edge. The monarch (*Danaus plexippus*), responded strongly only to tree-lined edges.

Patch shape is another important consideration in this fragmented system. Many remnant prairies occur along roadsides and railroad rights-of-way. Other linear habitats include the buffer and filter strips planted between crop fields and riparian systems to reduce erosion and the perennial strips included under Scenario 3 strip intercropping. Vegetation composition and structure in these linear habitats can elicit significant responses with respect to both the abundance and species richness of butterflies (Ries et al. 2001; Reeder et al. 2005). The width of the linear habitat may also influence butterfly species distribution patterns. Reeder et al. (2005) found that both the diversity and abundance of butterflies were positively correlated with the width of filter strips. This positive relationship was driven by habitat-restricted butterflies, most specifically the larger species in the Nymphalidae family such as *D. plexippus*, *S. idalia*, and red admiral (*Vanessa atalanta*). Conversely, richness of disturbance-tolerant butterflies was not correlated with strip width.

The species composition of linear prairie remnants often differs from larger block prairie remnants. Clausen et al. (2001) found that most butterfly species were significantly more abundant in nonlinear habitats, similar to the bioreserves in Scenario 3, than in linear habitats. Skibbe (2005), however, found both habitat-restricted and even threatened species existing in linear prairie remnants in Iowa. In some cases, these linear remnants may be geographic artifacts with rare species just barely hanging on in the last remaining patches. From a conservation perspective, then, in some cases, linear native habitats like the perennial strips and stream buffers in Scenario 3 may be highly valuable.

Linear habitats can also serve as connectors between fragmented prairie blocks. If habitat-restricted species could travel through linear habitats that were either remnants or planted to prairie vegetation, this movement might facilitate dispersal to native or restored prairie sites. For the baseline time period, the Iowa landscape did not have many linear prairie sites that actually connect to blocks of prairie. However, the biodiversity target zone under Scenario 3 is intended to facilitate native species dispersion through native perennial strip intercropping. Haddad and Baum (1999) found that habitat-restricted butterfly species reached higher densities in patches connected by corridors than in similar isolated patches. Sutcliffe et al. (2003) modeled the benefits of farmland restoration on butterfly movement and found that restoring grassy banks between arable fields significantly increased connectivity between subpopulations.

The configuration of prairie restorations within the larger landscape context can also influence butterfly diversity and abundance. Shepherd and Debinski (2005a) found that integrated reconstructions (i.e., planting units in a larger matrix of reconstructed and remnant prairies) have higher species richness than isolated reconstructions.

Finally, the combination of local effects and landscape context effects can have separate yet important influences on the distribution of butterfly species (Weibull and Ostman 2003; Bergman et al., 2004; Skibbe 2005, but see Collinge et al. 2003). For example, the swallowtails (Papilionidae) use forests especially when they are adjacent to riparian areas. In our approach, in which we ranked each habitat patch individually, we have probably underestimated the potential use that swallowtails might have for the wooded areas near streams that characterize landscape futures under both Scenarios 2 and 3.

In summary, many variables interact at multiple scales to influence distribution of pollinators across the landscape. The results described here provide one way to estimate the benefits of protecting wetlands, grasslands, savannas, and native prairies. We can predict how biodiversity of butterflies may respond to certain types of landscape change; for example, that Scenario 2 may be better for some species than Scenario 3. Intensive agriculture may have much more negative effects on Lepidopteran diversity in more topographically diverse regions such as Buck Creek. Habitat management, patch shape, patch adjacency, and landscape context will also influence butterfly diversity and abundance. In many cases, we would expect these effects to be positive, increasing the abundance and diversity of butterflies within the landscape. In some cases, though, if habitats exist, yet are too small or too widely dispersed, the probability of some of these rare species being present or maintaining small populations may be less than our modeling results would indicate.

## Implications for Agricultural Policy

Maintaining pollinator diversity in agricultural systems may be critically important to the health and sustainability of those systems, particularly as we enter a period of potentially rapid environmental change. The ecosystem services provided by pollinators are critical to agriculture, and in many cases, irreplaceable. We cannot afford to lose the pollinators of the crops of the future through shortsighted management practices that focus only on today's crops and conditions. The modeled results presented here, as well as the results of other recent research on pollinator diversity in agricultural systems, indicate the importance of maintaining diverse areas with heterogeneous habitats adjacent to one another. Including diverse assemblages of native plants in agricultural landscapes to serve as host plants for diverse insect species also appears to be essential. Landscape ecological considerations of patch configuration and connectivity among habitats, as well as management practices applied to areas broadly intended to help conserve plant and animal biodiversity will also be important in determining the success of efforts to maintain native pollinator diversity.

## Acknowledgments

We thank Scott Mahady and Dennis Schlict for review of the butterfly habitat suitability rankings.

## Chapter 10

# Amphibian Population Dynamics

Heather L. Rustigian, Mary V. Santelmann,
and Nathan H. Schumaker

Agriculture has contributed to loss of vertebrate biodiversity in many regions, including the U.S. Corn Belt (Erlich 1988; Risser 1988; Herkert 1991; Soulé 1991; Freemark 1995; McNeely et al. 1995). Amphibian populations, in particular, have experienced widespread and often inexplicable declines, range reductions, and extinctions (Blaustein and Wake 1990; Wake 1991). Amphibians are an important group to monitor and model for many reasons. They respond to both aquatic and terrestrial conditions, and are considered by many to be barometers of ecosystem health. They were historically an important component of the food web in the region. Among non-fish vertebrates, amphibians are the group with the greatest proportion of threatened and endangered species in Iowa. Of the 23 amphibian species known to occur in Iowa, 12 are declining, threatened, or endangered within the state (Christiansen 1981), leading to forecasts that without a reversal, less than one-third of Iowa's present amphibian and reptile fauna will remain in 50 years. Few attempts have been made, however, to model amphibian population dynamics (Gibbs 1993; Halley et al. 1996). Even fewer models have been spatially explicit or have taken landscape complementation (the proximity of two critical and largely independent habitats and the degree to which amphibians can move between them) and climatic variability into account (Vos and Stumpel 1995). The paucity of life history and dispersal data for most amphibian species makes model parameterization difficult (Doak and Mills 1994; Halley et al. 1996). Because many species require both aquatic and terrestrial habitats, they are limited to areas where there is both sufficient moisture for reproduction and survival and access to adjacent terrestrial habitats (Wilbur 1987; John-Alder and Morrin 1990). In the Midwest, climatic variability is also an important component of amphibian population dynamics. Northern prairie wetlands undergo a 10- to 20-year wet–dry cycle, resulting in two- to three-year droughts (Duvick and Blasing 1981; Karl and Riebsame 1984). Droughts have been linked to suppressed amphibian reproductive activity (Dodd 1993), decreased reproductive success (Seale 1982; Semlitsch 1983, 1987; Skelly 1995, 1996), localized extinctions, and population declines (Blair 1957). For these reasons, models that incorporate both climatic variability and landscape complementation are useful in simulating amphibian population dynamics (Dunning et al. 1992; Taylor et al. 1993; Pope et al. 2000).

Spatially explicit population models (SEPMs) differ from other landscape and de-mographic models because they incorporate movement of organisms among specific habitat patches across the landscape, linking animal demographic and dispersal data with landscape structure to predict changes in populations over time. In the study described in this chapter, we applied an SEPM to evaluate alternative futures land-scapes for their potential impacts on population dynamics of four amphibian species in the two Iowa study watersheds (Buck and Walnut Creek) described in Chapter 4. We included considerations of climatic variability and habitat complementation in our modeling work, with a goal of quantifying the consequences of alternative landscape futures for populations of four amphibian species. At the end of the chapter, we discuss the implications of this research for agricultural policy.

## Conducting the Study

We used a customized version of PATCH (Program to Assist in Tracking Critical Habitat; Schumaker 1998), a female-only, individual-based demographic model that incorpo-rates effects of fluctuations in both environmental conditions and demographic sto-chasticity. Data requirements of the model include raster landscape maps, habitat-use patterns, maximum densities, maximum dispersal distances, and estimates of survival and fecundity. Of the 19 amphibian species known to occur in central Iowa, adequate life history data were available for 4 species—the tiger salamander (*Ambystoma tigri-num*), the American toad (*Bufo americanus*), the gray treefrog (*Hyla versicolor*), and the western chorus frog (*Pseudacris triseriata*). We chose these four species for our model-ing work. For more detailed description of the model and parameters, see Rustigian et al. (2003). For detailed descriptions of the baseline and historical landscapes for the Walnut Creek and Buck Creek study watersheds and the future scenarios, see Chapter 4 and Color Figures 30–43 in this volume.

We compiled a species–habitat association matrix from published literature and local expert opinion. Estimates of maximum female breeding and nonbreeding densi-ties, maximum dispersal distances, and average survival and reproductive rates were also obtained from the literature. We ran simulations over a 100-year period for each study species and eight landscapes (three futures and a baseline for each of the two study watersheds), increasing populations until steady states were reached in the first 50 years. Next, we averaged output data derived from PATCH over years 50 to 100. These data included the following parameters: population size, number of breeding females, observed dispersal distances, and breeding site occupancy rates. We ranked the landscape futures against the baseline landscape according to three metrics: (1) the mean number of breeding females (MNBF); (2) the saturation of breeding sites (MNBF relative to carrying capacity, where carrying capacity is defined as the number of breed-ing sites multiplied by the maximum number of breeding females allowed to occupy a site); and (3) breeding site occupancy (percentage of breeding sites with at least one breeding female during years 50 to 100). To quantify the landscape structure of the futures by assessing the amount, quality, and spatial arrangement of breeding habitat, we used pattern-based metrics.

# Assessing the Alternatives

## Changes in Habitat Area

Comparing the areas in habitat of each alternative future, less than 1 percent of Walnut Creek watershed consisted of breeding habitat for the study species, except in Scenario 3 (Color Figure 43), where breeding habitat increased to just over 4 percent of the landscape. Under different scenarios, breeding habitat made up between 0.42 percent and 1.06 percent of Buck Creek watershed. Mean nearest neighbor distances for breeding habitat patches in Walnut Creek watershed were greatest in the baseline landscape (Color Figure 30) and smallest in Scenario 3 (Color Figure 43) for all species. In the Walnut Creek watershed, breeding habitat consisted primarily of roadside ditches (considered low- quality breeding habitat for the tiger salamander and gray treefrog) in the baseline landscape and in Scenarios 1 and 2. Under Scenario 3, it consisted primarily of ephemeral, seasonal, and semipermanent wetlands.

In Buck Creek watershed, the mean nearest neighbor distance for breeding habitat patches was greatest in the baseline landscape (Color Figure 34) for all species, and it was smallest in Scenario 1 (Color Figure 38) for the tiger salamander, gray treefrog, and American toad, and in Scenario 2 (Color Figure 40) for the western chorus frog. In Buck Creek, ponds made up a far larger percentage of breeding habitat than did roadside ditches.

## Amphibian Responses to Habitat under Different Scenarios

For Walnut Creek the MNBF showed little change relative to the baseline for Scenarios 1 and 2 for all species. In Scenario 3, where there was a nearly 100-fold increase in the amount of breeding habitat compared with the baseline, the MNBF increased from 3.5 (for the American toad) to 35 times greater than the baseline (for the tiger salamander; see Table 10-1).

The response in breeder saturation and breeding site occupancy to the scenarios varied among species in Walnut Creek (Table 10-1). Tiger salamander populations were at less than 20 percent of carrying capacity in the baseline and in Scenarios 1 and 2, but increased to more than 55 percent of carrying capacity in Scenario 3. This increase was caused by the Walnut Creek watershed wetland habitat restoration within the reserve under Scenario 3. Low salamander breeder saturation in Walnut Creek Scenarios 1 and 2, where breeding habitat is predominantly ditches, is a consequence of ditches being very poor sink habitats for this species. When we ran our models of Walnut Creek amphibian dynamics out beyond 50 years from the present under Scenarios 1 and 2, nearly 80 percent of Walnut Creek breeding sites for the tiger salamander were unoccupied, and most of these unoccupied sites were roadside ditches. Most of the occupied sites were large ponds clustered in the southeastern corner of the watershed, large temporary pools clustered in the northwest, and sections of roadside ditches near these ponds and pools (Color Figure 46a). The percentage of breeding sites occupied by tiger salamanders more than tripled in Scenario 3 for Walnut Creek watershed because a wetland reserve was added (Color Figures 15 and 46b).

Populations of the gray treefrog were near 70 percent of carrying capacity in the baseline and in Scenarios 2 and 3 in Walnut Creek, but were only 53 percent of carrying capacity under Scenario 1, possibly because of the slight increase in roadside

**Table 10-1.** Response of the Four Amphibian Species to Land Cover Changes under Alternative Scenarios

| Mean number of breeding females | | | |
|---|---|---|---|
| **Walnut Creek** | *Ambystoma tigrinum* | *Pseudacris triseriata* | *Hyla versicolor* | *Bufo americanus* |
| Baseline | 3,346.43 | 4,892.45 | 8,224.73 | 4,941.29 |
| Scenario 1 | 3,022.49 | 4,628.28 | 6,583.18 | 5,267.73 |
| Scenario 2 | 3,814.29 | 6,447.00 | 8,618.84 | 5,755.80 |
| Scenario 3 | 118,117.40 | 45,978.53 | 89,893.24 | 17,386.51 |
| **Buck Creek** | *Ambystoma tigrinum* | *Pseudacris triseriata* | *Hyla versicolor* | *Bufo americanus* |
| Baseline | 32,104.55 | 18,247.82 | 20,836.59 | 17,696.94 |
| Scenario 1 | 30,091.94 | 17,080.51 | 18,776.25 | 17,473.82 |
| Scenario 2 | 30,000.92 | 16,845.08 | 20,544.88 | 17,974.84 |
| Scenario 3 | 37,816.33 | 23,201.20 | 25,705.86 | 19,351.49 |

| Breeding site saturation | | | |
|---|---|---|---|
| **Walnut Creek** | *Ambystoma tigrinum* | *Pseudacris triseriata* | *Hyla versicolor* | *Bufo americanus* |
| Baseline | 0.165 | 0.402 | 0.676 | 0.783 |
| Scenario 1 | 0.147 | 0.375 | 0.533 | 0.817 |
| Scenario 2 | 0.183 | 0.472 | 0.69 | 0.829 |
| Scenario 3 | 0.552 | 0.357 | 0.70 | 0.381 |
| **Buck Creek** | *Ambystoma tigrinum* | *Pseudacris triseriata* | *Hyla versicolor* | *Bufo americanus* |
| Baseline | 0.87 | 0.824 | 0.941 | 0.992 |
| Scenario 1 | 0.852 | 0.806 | 0.886 | 0.991 |
| Scenario 2 | 0.829 | 0.742 | 0.765 | 0.986 |
| Scenario 3 | 0.845 | 0.864 | 0.957 | 0.992 |

| Breeding site occupancy | | | |
|---|---|---|---|
| **Walnut Creek** | *Ambystoma tigrinum* | *Pseudacris triseriata* | *Hyla versicolor* | *Bufo americanus* |
| Baseline | 21.6 | 87.3 | 93.6 | 98.8 |
| Scenario 1 | 21.2 | 79.8 | 73.2 | 98.9 |
| Scenario 2 | 28.0 | 83.6 | 96.2 | 99.1 |
| Scenario 3 | 92.0 | 98.6 | 99.0 | 97.2 |
| **Buck Creek** | *Ambystoma tigrinum* | *Pseudacris triseriata* | *Hyla versicolor* | *Bufo americanus* |
| Baseline | 90.7 | 85.6 | 98.1 | 100 |
| Scenario 1 | 91.8 | 85.0 | 93.6 | 100 |
| Scenario 2 | 90.9 | 80.7 | 97.3 | 100 |
| Scenario 3 | 90.7 | 90.9 | 98.7 | 100 |

ditches as breeding habitat. According to model results, the gray treefrog would be expected to occupy more than 90 percent of available breeding sites in a landscape such as the Walnut Creek baseline and Scenarios 2 and 3, but only 73 percent of breeding sites under Scenario 1 (Table 10-1).

Populations of the American toad were near 80 percent of carrying capacity in Walnut Creek baseline and in Scenarios 1 and 2, but only 38 percent of carrying capacity under Scenario 3. Low breeder saturation for the American toad in Scenario 3 can be attributed to the low reproductive rate of the toad compared to the other species, and the resulting slow increase of toad populations following population declines in dry years. The American toad occupied more than 97 percent of available breeding sites in all Walnut Creek landscape futures (Breeding site occupancy in Table 10-1).[1]

Populations of the western chorus frog were near 40 percent of carrying capacity in all Walnut Creek landscape futures (Table 10-1). Although this species occupied more than 75 percent of the breeding sites in these landscapes, the percentage of sites occupied was lowest under Scenario 1.

Relative to the baseline landscape, the MNBF showed less variation among landscape futures for all species in Buck Creek than in Walnut Creek (Table 10-1). As in Walnut Creek, Buck Creek Scenario 3 ranked highest in MNBF for each species, with the mean number of breeders ranging from 1.09 (for the American toad) to 1.27 (for the western chorus frog) times the MNBF in the baseline landscape.

Buck Creek breeder saturation for the tiger salamander and the American toad showed little variation among landscapes, with populations at more than 82 and 98 percent of carrying capacity, respectively. Breeding site occupancy in Buck Creek showed little variation among landscapes for the tiger salamander, the gray treefrog, and the American toad, with more than 90 percent of sites occupied in every simulation (Table 10-1). Western chorus frog breeder saturation (74.2 percent of carrying capacity), and breeding site occupancy (81 percent of sites occupied) were lowest in Scenario 2 (Color Figure 40), possibly because of infrequent utilization of the engineered wetlands in Scenario 2. Gray treefrog breeder saturation was also lowest in Scenario 2, with populations at less than 80 percent of carrying capacity.

## Influence of Alternative Futures on Amphibian Populations

Gray treefrog and American toad populations were predicted by the model to be at more than 70 percent carrying capacity in the baseline landscape in both watersheds. For the baseline Buck Creek landscape, tiger salamander and western chorus frog populations were predicted to be at more than 70 percent of carrying capacity, and for the baseline Walnut Creek landscape, the populations of the tiger salamander and the western chorus frog were predicted to be at less than 50 percent of carrying capacity. These relatively high percentages suggest that all of these species may be at risk as landscapes change.

Scenario 3 consistently ranked above the baseline and the other scenarios in mean number of breeders supported, but patterns in the other responses we investigated were less clear (Color Figures 42 and 43). The obvious advantages of Scenario 3 were the increase in the amount of breeding habitat; the decrease in proportion of breed-

---

1.   The reason there is low breeder saturation even though there is high breeding site occupancy is that the number of breeding females in the breeding site is below carrying capacity. Although most of the sites are occupied, the breeding sites have fewer females than they are capable of holding because of low intrinsic reproductive rates for the toad.

ing habitat as roadside ditches (which function as sinks for two of the study species); and the reduced isolation of breeding sites. In contrast, Scenarios 1 and 2 demonstrated little benefit for amphibian species compared to the baseline landscape (Color Figures 38-41). Landscape modifications in Scenario 2, such as increased land area in forest and pasture, wider riparian buffers, and the creation of engineered wetlands, did little to affect amphibian viability. Only Scenario 3 showed substantial increases in the amount of high-quality breeding habitat.

The model used in this study serves as a much-needed method to predict effects of landscape change on pond-breeding amphibian populations in agricultural regions. The four species modeled are all widespread generalists assigned a low conservation priority, and, assuming that parameter estimates used in the simulations are reasonably correct, our modeled results suggest that all four species will persist under baseline and all three future scenario conditions. However, their habitats, wetlands, and ponds are increasingly endangered habitats in the agricultural landscapes of the Midwest (Knutson et al. 2004), and this study demonstrated dramatic differences in the habitat characteristics of the alternative future scenarios with Scenario 3 providing much greater habitat gains for amphibians.

## Implications for Agricultural Policy

An increasing number of studies are evaluating the effect of agriculture on amphibians. Although Knutson et al. (1999) found no negative association between frog and toad presence and agriculture, Hecnar (1997) found decreased amphibian diversity associated with agricultural landscapes in Ontario, and Bonin et al. (1997) found an inverse relationship between anuran species richness and agricultural monocultures in Quebec. Furthermore, several studies have shown the toxic effects of pesticides and herbicides on amphibians (Berrill et al. 1997; Diana and Beasley 1998; Relyea 2005). These studies suggest that such agents play a role in the high prevalence of hind limb deformities (Ouellet et al. 1997; Helgen et al. 1998). Some agricultural practices (such as haying or grazing) may be more favorable to amphibians than others (such as row cropping), but no studies have been conducted to support such theories (Knutson et al. 1999). Field-based studies to quantify amphibian use of different types of agricultural lands would improve our level of knowledge about the impacts of agricultural practices and improve the accuracy of population viability models.

Amphibian habitat in agricultural landscapes is becoming increasingly fragmented (Knutson et al. 2004), and in habitats that do remain, substantial amphibian mortality may be resulting from exposure to herbicides and pesticides (Relyea 2005), including exposure to components once considered inert (such as surfactants). Both contamination of shallow wetlands from overspray and direct spraying of ditches to control weeds can result in unintended exposure of amphibians to potentially toxic agricultural chemicals. Practices used to manage weeds in ditches may be a serious concern, and planting roadsides with native vegetation to discourage weeds and controlling vegetation height in ditches by mowing instead of using herbicides may help protect amphibian populations. In addition, critical breeding sites must be protected from overspray by ensuring that at least some wetlands and ponds are shielded from agrichemical applications. For example, creation of prairie-wetland reserves in which breeding ponds occur in a matrix of native prairie vegetation, such as those in Scenario 3 (Color Figure 15), may ensure that currently common and widespread amphibian species retain that status.

## Acknowledgments

We thank Denis White for assistance with geographic information systems (GIS), and Michael Lannoo for reviewing the species habitat matrix. Funding was provided by the U.S. Environmental Protection Agency (U.S. EPA) STAR Grants (Water and Watersheds) program, #R825335-01-0. The information in this document was funded in part by the U.S. EPA, and it has been subjected to the agency's peer and administrative review and approved for publication. The conclusions and opinions are solely those of the authors and are not necessarily the views of the agency.

## Chapter 11

# Impacts on Mammal Communities: A Spatially Explicit Model

Mark E. Clark, Brent J. Danielson, Mary V. Santelmann,
Joan Iverson Nassauer, Denis White, and
Kathryn Freemark Lindsay

In this chapter, we link a metapopulation model to multiple-habitat landscapes to simulate the response of mammal communities to changes in land use and management for the two Iowa study watersheds (see Chapter 4). Although many authors have similarly modeled individual species in complex landscapes, here we attempt to model an entire mammalian community with simultaneous interactions among species. This recognizes that species do not exist as independent entities. Even though this is a well-known fact to ecologists, we are unaware of any conservation planning or policy-making that considers the full complexity of entire taxonomic communities. For both theoretical and empirical reasons, we believe that such complexity is both far-reaching and not easily intuited from simpler models, and more complex models may even produce dynamics that are completely contrary to expectation (e.g., Sih et al. 1985; Stone and Roberts 1992; Bascompte et al. 2006). For example, a predator may sometimes act to increase the abundance of some prey species, or two species competing for resources may act as mutualists in the presence of a third or fourth species, all as a result of the complex arrays of indirect effects among the diverse communities that exist in any landscape. Consequently, landscape models intended to inform biodiversity conservation policy may find quite different results when the response of an entire community is considered, rather than that of a single species in isolation.

Here, we use the results of modeling simulations to quantitatively compare alternative future landscapes for the three scenarios (see Chapter 4) to a baseline landscape. We also develop several indices to quantify diversity, population densities, and viability in the mammal community and to facilitate normalized comparisons among landscapes. Multiple indices were necessary to evaluate the effects of proposed landscape changes in the watersheds because species interactions often lead to trade-offs between species diversity and population densities.

## Conducting the Study

The model we used consists of a set of metapopulations in a landscape of polygon habitat patches in which patch- and species-specific population models are linked via immigration and emigration. State variables (such as population mortality or dispersal among populations) for these species-specific population models depend on both species life history traits and patch characteristics. For example, dispersal (immigration and emigration) between populations depends on species characteristics as well as the relative location of habitat patches for that species within the landscape. Species interactions were incorporated into the patch population models in the form of Lotka-Volterra mass-action dynamics for competitors and Holling type-2 dynamics for predators and prey. We developed population models for each of 52 mammal species that could potentially inhabit the study area (Bowles 1975).

### The Watersheds and Future Scenarios

Chapter 4 describes the two study watersheds and the alternative scenarios and landscape futures in detail. In our simulations, the landscape futures are represented as collections of contiguous polygons of uniform land cover. Land cover was classified using 27 categories (Table 11-1), and remains static through all simulations.

From the baseline data, three alternative landscape futures were designed for each watershed (Chapter 4). These alternative futures reflect potential landscape changes associated with the policy priorities of three different scenarios.

### Metapopulation Model Overview

Discrete-time, density-dependent populations (patches) are connected via dispersal to simulate a metapopulation in the landscape. As a result, for a population in a given patch in the landscape, the density of the population at time $t+1$, $N_{t+1}$ (number females/hectare), is given by

$$N_{t+1} = f(N_t) - E + I \qquad (11\text{-}1)$$

where $f(N)$ is a population growth function, $E$ is the density exiting the patch to other patches in the landscape, and $I$ is the density immigrating into the patch from other patches. We use a two-parameter function developed by May (1975) for density-dependent population growth in which

$$f(N) = N \cdot e^{r \cdot (1 - \frac{2 \cdot N}{K})}. \qquad (11\text{-}2)$$

This function exhibits stable, cyclic, and chaotic behavior when $r < 2$, $2 \leq r < 3$, and $r \geq 3$, respectively. The parameters of Equation 11-2 are represented by the intrinsic rate of growth ($r$, unitless) and carrying capacity ($K$, number/hectare). In the model, both $r$ and $K$ depend on patch habitat type and the life history of each species. A proportion of the population is assumed to migrate to other patches based on species-specific habitat preferences, and their survival depends on the distance they move and on species-specific maximum dispersal distances ($D_{max}$, kilometers).

We developed a dynamic community model by adding Lotka-Volterra species interactions into Equation 11-1. Interaction between a species and a competitor ($C$, number of females/hectare) in the same patch is assumed to follow a simple mass

**Table 11-1.**   Land Cover Categories Used to Classify Habitat Patches[a]

| Category | Land Cover Type | Description |
|---|---|---|
| 1 | Row crop | Corn or soybean crop using chisel plowing |
| 2 | No-till crop | Corn or soybean crop using conservation tillage |
| 3 | Strip intercrop | Alternating strips of corn, soybean, and native crops |
| 4 | Small grains | Oat, barley, or wheat crops |
| 5 | Fallow land | Uncropped field treated with herbicide |
| 6 | Organic strip | Intercrop strips of organic corn, beans, and alfalfa |
| 7 | Organic crop | Organic corn or soybean monocultures |
| 8 | Farmstead | Farmstead or urban area |
| 9 | Herbaceous strip | Fencerows, roadsides, and field borders with herbaceous vegetation |
| 10 | Woody strip | Fencerows, roadsides, and field borders with woody vegetation |
| 11 | Conservation Reserve Program (CRP) | Ungrazed grasslands of brome or prairie grasses; Perennial non-crop |
| 12 | Hay | Mowed alfalfa or grasses |
| 13 | Dry prairie | Upland tallgrass prairie |
| 14 | Wet prairie | Sedge meadows and potholes |
| 15 | Pasture | Grazed areas with small (<16.40 feet (5 m)) trees and grasses |
| 16 | Shrubland | Ungrazed areas with small (<16.40 feet (5 m)) trees and grasses |
| 17 | Ungrazed forest | Ungrazed oak–hickory–basswood forest with 25–75 percent canopy closure and <50-year-old stands |
| 18 | Ungrazed floodplain | Ungrazed bottomland hardwood forest (maple, cottonwood, and ash) with 25–75 percent canopy closure |
| 19 | Ungrazed upland forest | Ungrazed oak–hickory–basswood forest with >75 percent canopy closure and >50-year-old stands |
| 20 | Grazed upland forest | Grazed oak–hickory–basswood forest with >75 percent canopy closure and >50-year-old stands |
| 21 | Ungrazed savanna | Ungrazed grassland with scattered trees, <25 percent canopy closure |
| 22 | Grazed savanna | Grazed grassland with scattered trees, <25 percent canopy closure |
| 23 | Wetland | Semipermanent, seasonally wet prairie pothole |
| 24 | Pond | Small farm ponds with permanent water |
| 25 | Stream | Permanent stream channel |
| 26 | Engineered wetland | Detention ponds |
| 27 | Road | Paved and gravel roads |

[a] Patches were classified in the geographic information systems (GIS) landscape coverage for the spatial component of the metapopulation model.

action function of densities given by $g(N,C) = \alpha \cdot N \cdot C$. Interaction between a species and a predator or prey ($P$, number of females/hectare) in the same patch is assumed to follow a Holling type-2 functional response, in which $h(N,P) = \beta \cdot \frac{N \cdot P}{(1+P)}$ if $P$ preys on $N$ or $h(N,P) = \beta \cdot \frac{N \cdot P}{(1+N)}$ if $N$ preys on $P$. Adding the species interaction terms, Equation 11-1 for the density of a species within a patch becomes

$$N_{t+1} = f(N_t) + \sum_{\text{all competitors}} g(N,C) + \sum_{\text{all predators/prey}} h(N,P) - E + I. \qquad (11\text{-}3)$$

Summing the product of population density given in Equation 11-3 and patch area across all patches in the landscape yields the abundance of each species in the landscape. Detailed equations for computing the population dynamics of the community metapopulation can be obtained from the lead author.[1] The 52 species potentially inhabiting the landscape are listed in Table 11-2.

Where possible, we obtained estimates for the key model parameters, population growth rate ($r$), carrying capacity ($K$), and maximum dispersal distance ($D_{max}$, kilometers), and habitat suitability ratings for 52 mammals from the literature (Bowles 1975). We identified these 52 mammals as potential inhabitants of watersheds in central Iowa. When estimates were unavailable in the literature, we used allometric relationships (i.e., functions based on body size) between life history data to estimate values. We used maximum dispersal distance for each species ($D_{max}$) in determining immigration and emigration among patches. Habitat suitability ratings (shown in Table 11-2) were used to adjust life history parameters for the different habitats. Table 11-3 gives the values of the parameters we used in model simulations. Values for parameters in the mass action competition function and Holling type-2 predator-prey functional response for all pairwise combinations of species, as well as calculations and description of model parameters can be obtained from the lead author on request.

### Indices for Comparison

We computed several indices for diversity and density from simulation results to facilitate comparison among landscapes and among species. We used two measures to evaluate diversity. First, we computed raw species diversity (richness) as the number of breeding species (i.e., species with at least two individuals present at the end of the 100-year simulation). Second, we used relative diversity—computed as the ratio of number of breeding species based on simulation results and the number of potentially breeding species based on habitat patch carrying capacities—to measure the efficiency of a given landscape to maintain total community diversity. Both diversity indices can be computed across the whole landscape to compare among different landscapes, and within a landscape to compare among habitat types.

We used two measures to evaluate density. First, we computed mean population density (number of individuals per hectare) over the last 90 years (years 1 through 10 were ignored to eliminate bias from initial conditions) of the simulation. Second, we used relative density (ratio of the mean population density predicted in simulation and the population density from habitat patch carrying capacities), to measure the efficiency at which individual species exploited the landscape. Again, both density indices can be used to compare across landscapes and within landscapes, but these indices can also be used to compare across species.

---

1.  Mark Clark, Department of Biological Sciences, North Dakota State University, Fargo, North Dakota, 58105. Email: m.e.clark@ndsu.edu.

**Table 11-2.** List of Species and Habitat Ratings by Land Cover Category

| Species | Land cover category | | | | | | | | | | | | | | | | | | | | | | | | | | |
|---|---|---|---|---|---|---|---|---|---|---|---|---|---|---|---|---|---|---|---|---|---|---|---|---|---|---|---|
| | 1 | 2 | 3 | 4 | 5 | 6 | 7 | 8 | 9 | 10 | 11 | 12 | 13 | 14 | 15 | 16 | 17 | 18 | 19 | 20 | 21 | 22 | 23 | 24 | 25 | 26 | 27 |
| *Didelphis virginiana* | 2 | 2 | 2 | 2 | 2 | 2 | 2 | 4 | 1 | 3 | 2 | 2 | 2 | 2 | 1 | 3 | 4 | 4 | 4 | 3 | 3 | 2 | 0 | 0 | 0 | 0 | 0 |
| *Sorex cinereus* | 0 | 1 | 2 | 0 | 1 | 2 | 0 | 0 | 2 | 3 | 3 | 1 | 2 | 3 | 1 | 2 | 3 | 3 | 2 | 1 | 1 | 0 | 0 | 0 | 0 | 0 | 0 |
| *Blarina brevicauda* | 1 | 3 | 3 | 0 | 2 | 3 | 0 | 2 | 3 | 2 | 4 | 3 | 3 | 4 | 3 | 4 | 3 | 3 | 3 | 1 | 4 | 2 | 0 | 0 | 0 | 0 | 0 |
| *Cryptotis parva* | 0 | 2 | 3 | 0 | 3 | 3 | 0 | 1 | 3 | 2 | 4 | 3 | 4 | 3 | 1 | 3 | 0 | 0 | 0 | 0 | 4 | 2 | 0 | 0 | 0 | 0 | 0 |
| *Scalopus aquaticus* | 0 | 1 | 2 | 0 | 1 | 2 | 0 | 3 | 2 | 2 | 3 | 2 | 3 | 3 | 2 | 4 | 4 | 3 | 4 | 3 | 3 | 2 | 0 | 0 | 0 | 0 | 0 |
| *Myotis lucifugus* | 2 | 3 | 3 | 2 | 3 | 3 | 3 | 4 | 3 | 3 | 3 | 2 | 3 | 3 | 2 | 2 | 2 | 3 | 3 | 3 | 1 | 1 | 2 | 4 | 4 | 2 | 0 |
| *Myotis sodalis* | 2 | 3 | 3 | 2 | 3 | 3 | 3 | 0 | 3 | 2 | 3 | 2 | 2 | 3 | 2 | 2 | 2 | 4 | 3 | 2 | 2 | 2 | 2 | 4 | 4 | 2 | 0 |
| *Myotis septentrionalis* | 2 | 3 | 3 | 2 | 3 | 3 | 3 | 2 | 3 | 3 | 3 | 2 | 2 | 3 | 2 | 2 | 4 | 4 | 2 | 1 | 1 | 1 | 2 | 4 | 4 | 2 | 0 |
| *Lasionycteris noctivagans* | 2 | 3 | 3 | 2 | 3 | 3 | 3 | 3 | 3 | 3 | 3 | 2 | 2 | 3 | 2 | 2 | 4 | 4 | 3 | 3 | 2 | 2 | 2 | 4 | 4 | 2 | 0 |
| *Pipistrellus subflavus* | 2 | 3 | 3 | 2 | 3 | 3 | 3 | 3 | 3 | 3 | 3 | 2 | 2 | 3 | 3 | 2 | 2 | 2 | 3 | 3 | 3 | 3 | 2 | 4 | 4 | 2 | 0 |
| *Eptesicus fuscus* | 2 | 3 | 3 | 2 | 3 | 3 | 3 | 4 | 3 | 2 | 2 | 2 | 2 | 3 | 2 | 2 | 3 | 4 | 4 | 3 | 3 | 1 | 2 | 4 | 4 | 2 | 0 |
| *Lasiurus borealis* | 2 | 3 | 3 | 2 | 3 | 3 | 3 | 2 | 3 | 3 | 3 | 2 | 2 | 3 | 2 | 2 | 4 | 4 | 3 | 3 | 1 | 1 | 2 | 4 | 4 | 2 | 0 |
| *Lasiurus cinereus* | 2 | 3 | 3 | 2 | 3 | 3 | 3 | 2 | 3 | 2 | 3 | 2 | 2 | 3 | 2 | 2 | 4 | 4 | 4 | 4 | 1 | 1 | 2 | 4 | 4 | 2 | 0 |
| *Nycticeius humeralis* | 2 | 3 | 3 | 2 | 3 | 3 | 3 | 4 | 3 | 4 | 3 | 2 | 2 | 3 | 2 | 3 | 3 | 4 | 4 | 4 | 2 | 2 | 2 | 4 | 4 | 0 | 0 |
| *Sylvilagus floridanus* | 2 | 3 | 3 | 1 | 3 | 3 | 1 | 3 | 3 | 2 | 3 | 1 | 3 | 2 | 3 | 4 | 3 | 3 | 3 | 2 | 3 | 3 | 0 | 0 | 0 | 0 | 0 |
| *Lepus townsendii* | 3 | 3 | 2 | 3 | 3 | 2 | 3 | 0 | 2 | 0 | 0 | 3 | 3 | 1 | 2 | 0 | 0 | 0 | 0 | 0 | 1 | 2 | 0 | 0 | 0 | 0 | 0 |
| *Tamias striatus* | 0 | 0 | 0 | 0 | 0 | 0 | 0 | 4 | 0 | 3 | 0 | 0 | 0 | 0 | 3 | 2 | 4 | 4 | 4 | 4 | 1 | 1 | 0 | 0 | 0 | 0 | 0 |
| *Marmota monax* | 0 | 0 | 2 | 0 | 1 | 0 | 0 | 2 | 3 | 3 | 4 | 3 | 4 | 2 | 3 | 3 | 2 | 1 | 2 | 2 | 4 | 3 | 3 | 0 | 0 | 0 | 0 |
| *Spermophilus tridecemlineatus* | 0 | 0 | 3 | 1 | 2 | 0 | 0 | 3 | 4 | 0 | 4 | 3 | 4 | 3 | 2 | 1 | 0 | 0 | 0 | 0 | 2 | 1 | 0 | 0 | 0 | 0 | 0 |
| *Spermophilus franklinii* | 0 | 0 | 0 | 0 | 0 | 0 | 0 | 0 | 3 | 0 | 3 | 1 | 4 | 3 | 0 | 0 | 0 | 0 | 0 | 0 | 3 | 1 | 0 | 0 | 0 | 0 | 0 |
| *Sciurus carolinensis* | 0 | 0 | 0 | 0 | 0 | 0 | 0 | 4 | 0 | 2 | 0 | 0 | 0 | 0 | 0 | 0 | 2 | 2 | 4 | 3 | 0 | 0 | 0 | 0 | 0 | 0 | 0 |

**Table 11-2.** List of Species and Habitat Ratings by Land Cover Category (continued)

| Species | \- | \- | \- | \- | \- | \- | \- | \- | \- | Land cover category | | | | | | | | | | | | | | | | | |
|---|---|---|---|---|---|---|---|---|---|---|---|---|---|---|---|---|---|---|---|---|---|---|---|---|---|---|---|
| | 1 | 2 | 3 | 4 | 5 | 6 | 7 | 8 | 9 | 10 | 11 | 12 | 13 | 14 | 15 | 16 | 17 | 18 | 19 | 20 | 21 | 22 | 23 | 24 | 25 | 26 | 27 |
| *Sciurus niger* | 0 | 0 | 0 | 0 | 0 | 0 | 0 | 4 | 0 | 3 | 0 | 0 | 0 | 0 | 1 | 1 | 3 | 4 | 4 | 4 | 1 | 2 | 0 | 0 | 0 | 0 | 0 |
| *Glaucomys volans* | 0 | 0 | 0 | 0 | 0 | 0 | 0 | 3 | 0 | 0 | 0 | 0 | 0 | 0 | 0 | 0 | 2 | 4 | 4 | 3 | 1 | 1 | 0 | 0 | 0 | 0 | 0 |
| *Geomys bursarius* | 0 | 0 | 2 | 0 | 1 | 2 | 0 | 1 | 1 | 0 | 0 | 1 | 3 | 1 | 1 | 2 | 0 | 0 | 0 | 0 | 3 | 1 | 0 | 0 | 0 | 0 | 0 |
| *Perognathus flavescens* | 2 | 2 | 2 | 2 | 2 | 2 | 2 | 1 | 1 | 1 | 1 | 2 | 2 | 1 | 2 | 0 | 0 | 0 | 0 | 0 | 0 | 2 | 0 | 0 | 0 | 0 | 0 |
| *Castor canadensis* | 0 | 0 | 0 | 0 | 0 | 0 | 0 | 0 | 0 | 0 | 0 | 0 | 0 | 0 | 0 | 0 | 0 | 0 | 0 | 0 | 0 | 0 | 0 | 1 | 4 | 1 | 0 |
| *Reithrodontomys megalotis* | 0 | 1 | 3 | 0 | 2 | 3 | 0 | 0 | 3 | 0 | 4 | 2 | 4 | 2 | 1 | 2 | 0 | 0 | 3 | 0 | 3 | 1 | 0 | 0 | 0 | 0 | 0 |
| *Peromyscus maniculatus* | 2 | 3 | 4 | 2 | 3 | 4 | 0 | 1 | 3 | 0 | 4 | 2 | 4 | 3 | 2 | 1 | 0 | 0 | 0 | 0 | 2 | 2 | 0 | 0 | 0 | 0 | 0 |
| *Peromyscus leucopus* | 1 | 1 | 1 | 0 | 0 | 1 | 0 | 3 | 1 | 4 | 1 | 0 | 1 | 0 | 3 | 4 | 4 | 4 | 4 | 4 | 2 | 2 | 0 | 0 | 0 | 0 | 0 |
| *Onychomys leucogaster* | 2 | 3 | 2 | 2 | 3 | 2 | 2 | 0 | 0 | 0 | 0 | 1 | 0 | 1 | 0 | 0 | 0 | 0 | 0 | 0 | 3 | 2 | 0 | 1 | 0 | 0 | 0 |
| *Microtus pennsylvanicus* | 0 | 0 | 3 | 0 | 1 | 3 | 0 | 4 | 4 | 4 | 4 | 2 | 3 | 4 | 1 | 2 | 0 | 0 | 0 | 0 | 3 | 0 | 0 | 0 | 0 | 0 | 0 |
| *Microtus ochrogaster* | 0 | 0 | 3 | 0 | 1 | 3 | 0 | 0 | 4 | 0 | 4 | 2 | 4 | 3 | 1 | 2 | 0 | 0 | 3 | 0 | 3 | 0 | 0 | 0 | 0 | 0 | 0 |
| *Microtus pinetorum* | 0 | 0 | 0 | 0 | 0 | 0 | 0 | 1 | 0 | 2 | 1 | 0 | 1 | 1 | 2 | 2 | 3 | 3 | 3 | 2 | 1 | 0 | 0 | 0 | 0 | 0 | 0 |
| *Ondatra zibethicus* | 0 | 0 | 0 | 0 | 0 | 0 | 0 | 0 | 0 | 0 | 0 | 0 | 0 | 0 | 0 | 0 | 0 | 0 | 0 | 0 | 0 | 0 | 1 | 4 | 4 | 1 | 0 |
| *Synaptomys cooperi* | 0 | 0 | 2 | 0 | 0 | 2 | 0 | 0 | 3 | 1 | 3 | 1 | 3 | 3 | 0 | 3 | 0 | 0 | 0 | 3 | 0 | 0 | 0 | 0 | 0 | 0 | 0 |
| *Rattus norvegicus* | 1 | 2 | 1 | 1 | 1 | 1 | 0 | 4 | 1 | 2 | 0 | 0 | 0 | 0 | 0 | 0 | 0 | 0 | 0 | 0 | 0 | 0 | 0 | 0 | 0 | 0 | 0 |
| *Mus musculus* | 1 | 2 | 1 | 1 | 1 | 1 | 0 | 4 | 1 | 0 | 1 | 1 | 1 | 1 | 1 | 1 | 0 | 0 | 0 | 0 | 0 | 0 | 0 | 0 | 0 | 0 | 0 |
| *Zapus hudsonius* | 0 | 1 | 3 | 0 | 2 | 3 | 1 | 0 | 3 | 0 | 4 | 2 | 4 | 4 | 0 | 1 | 0 | 0 | 0 | 3 | 0 | 0 | 0 | 0 | 0 | 0 | 0 |
| *Canis latrans* | 0 | 2 | 3 | 0 | 1 | 3 | 1 | 1 | 2 | 2 | 3 | 1 | 3 | 3 | 3 | 4 | 3 | 3 | 4 | 3 | 3 | 3 | 0 | 0 | 0 | 0 | 0 |
| *Vulpes vulpes* | 1 | 2 | 3 | 0 | 1 | 3 | 1 | 3 | 2 | 3 | 2 | 2 | 3 | 3 | 3 | 4 | 2 | 2 | 2 | 3 | 3 | 3 | 0 | 0 | 0 | 0 | 0 |
| *Urocyon cinereoargenteus* | 1 | 1 | 2 | 0 | 1 | 2 | 2 | 2 | 1 | 1 | 1 | 1 | 1 | 1 | 1 | 2 | 3 | 3 | 3 | 2 | 1 | 1 | 0 | 0 | 0 | 0 | 0 |
| *Procyon lotor* | 1 | 2 | 3 | 1 | 1 | 3 | 0 | 4 | 2 | 2 | 2 | 2 | 2 | 2 | 2 | 3 | 4 | 4 | 4 | 3 | 2 | 2 | 1 | 0 | 0 | 0 | 0 |
| *Mustela erminea* | 0 | 1 | 3 | 0 | 1 | 3 | 0 | 1 | 2 | 1 | 3 | 1 | 3 | 2 | 2 | 3 | 0 | 0 | 0 | 3 | 2 | 2 | 0 | 0 | 0 | 0 | 0 |

**Table 11-2.** List of Species and Habitat Ratings by Land Cover Category (continued)

| Species | Land cover category | | | | | | | | | | | | | | | | | | | | | | | | | | |
|---|---|---|---|---|---|---|---|---|---|---|---|---|---|---|---|---|---|---|---|---|---|---|---|---|---|---|---|
| | 1 | 2 | 3 | 4 | 5 | 6 | 7 | 8 | 9 | 10 | 11 | 12 | 13 | 14 | 15 | 16 | 17 | 18 | 19 | 20 | 21 | 22 | 23 | 24 | 25 | 26 | 27 |
| *Mustela nivalis* | 0 | 1 | 3 | 0 | 1 | 3 | 0 | 1 | 2 | 1 | 3 | 1 | 3 | 4 | 2 | 3 | 0 | 0 | 0 | 0 | 3 | 2 | 0 | 0 | 0 | 0 | 0 |
| *Mustela frenata* | 0 | 1 | 2 | 0 | 1 | 2 | 0 | 1 | 1 | 3 | 2 | 0 | 2 | 2 | 2 | 4 | 3 | 3 | 3 | 2 | 3 | 2 | 0 | 0 | 0 | 0 | 0 |
| *Mustela vison* | 0 | 0 | 1 | 0 | 0 | 1 | 0 | 0 | 2 | 2 | 1 | 0 | 1 | 2 | 1 | 2 | 2 | 4 | 2 | 1 | 1 | 1 | 2 | 3 | 4 | 2 | 0 |
| *Taxidea taxus* | 0 | 1 | 3 | 1 | 1 | 3 | 1 | 1 | 2 | 1 | 4 | 2 | 4 | 3 | 1 | 3 | 1 | 1 | 1 | 1 | 3 | 2 | 1 | 0 | 0 | 0 | 0 |
| *Spilogale putorius inter-rupta* | 1 | 2 | 4 | 1 | 2 | 4 | 2 | 3 | 2 | 3 | 2 | 2 | 3 | 2 | 1 | 3 | 1 | 1 | 0 | 0 | 3 | 1 | 0 | 0 | 0 | 0 | 0 |
| *Mephitis mephitis* | 1 | 2 | 4 | 1 | 2 | 4 | 2 | 4 | 3 | 3 | 3 | 2 | 3 | 3 | 2 | 4 | 4 | 4 | 4 | 3 | 3 | 3 | 0 | 0 | 0 | 0 | 0 |
| *Odocoileus virginianus* | 3 | 3 | 4 | 1 | 1 | 4 | 1 | 1 | 3 | 3 | 3 | 1 | 2 | 2 | 3 | 4 | 4 | 3 | 3 | 4 | 4 | 3 | 0 | 0 | 0 | 0 | 0 |
| *Lynx rufus* | 0 | 1 | 1 | 0 | 1 | 1 | 0 | 0 | 1 | 1 | 2 | 0 | 2 | 2 | 1 | 2 | 2 | 3 | 3 | 1 | 2 | 1 | 1 | 0 | 0 | 0 | 0 |
| *Lutra canadensis* | 0 | 0 | 0 | 0 | 0 | 0 | 0 | 0 | 1 | 1 | 0 | 0 | 0 | 0 | 0 | 0 | 0 | 2 | 0 | 0 | 0 | 0 | 0 | 1 | 2 | 0 | 0 |

*Notes:* Habitat rated 0 is unusable by the species (*r* and *K* are set to 0). Habitat rated 1 is very poor (*r* and *K* are set to 0.01 and 0.1 of base value, respectively); habitat rated 2 is poor (*r* and *K* set to 0.1 and 0.2 of base values); habitat rated 3 is good (*r* and *K* both set to 0.7 of base value); and habitat rated 4 is the highest quality habitat and *r* and *K* are set to the base values (i.e., maximum).

**Key to Common names:** *Didelphis virginiana* (Virginia opossum); *Sorex cinereus* (Masked shrew); *Blarina brevicauda* (Northern Short-tailed shrew); *Cryptotis parva* (Least shrew); *Scalopus aquaticus* (Eastern mole); *Myotis lucifugus* (Little brown bat); *Myotis sodalist* (Indiana bat); *Myotis septentrionalis* (Northern myotis); *Lasionycteris noctivagans* (Silver-haired bat); *Pipistrellus subflavus* (Eastern pipistrelle); *Eptesicus fuscus* (Big brown bat); *Lasiurus borealis* (Red bat); *Lasiurus cinereus* (Hoary bat); *Nycticeius humeralis* (Evening bat); *Sylvilagus floridanus* (Eastern cottontail); *Lepus townsendii* (White-tailed jackrabbit); *Tamias striatus* (Eastern chipmunk); *Marmota monax* (Woodchuck); *Spermophilus tridecemlineatus* (Thirteen-lined ground squirrel); *Spermophilus franklinii* (Franklin's ground squirrel); *Sciurus carolinensis* (Gray squirrel); *Sciurus niger* (Fox squirrel); *Glaucomys volans* (Flying squirrel); *Geomys bursarius* (Plains pocket gopher); *Perognathus flavescens* (Plains pocket mouse); *Castor canadensis* (Beaver); *Reithrodontomys megalotis* (Western harvest mouse); *Peromyscus maniculatus* (Deer mouse); *Peromyscus leucopus* (White-footed mouse); *Onychomys leucogaster* (Northern grasshopper mouse); *Microtus pennsylvanicus* (Meadow vole); *Microtus ochrogaster* (Prairie vole); *Microtus pinetorum* (Woodland vole); *Ondatra zibethicus* (Muskrat); *Synaptomys cooperi* (Southern bog lemming); *Rattus norvegicus* (Norway rat); *Mus musculus* (House mouse); *Zapus hudsonius* (Meadow jumping mouse); *Canis latrans* (Coyote); *Vulpes vulpes* (Red fox); *Urocyon cinereoargenteus* (Gray fox); *Procyon lotor* (Raccoon); *Mustela erminea* (Ermine); *Mustela nivalis* (Least weasel); *Mustela frenata* (Long-tailed weasel); *Mustela vison* (Mink); *Taxidea taxus* (American badger); *Spilogale putorius interrupta* (Spotted skunk); *Mephitis mephitis* (Striped skunk); *Odocoileus virginianus* (White-tailed deer); *Lynx rufus* (Bobcat); *Lutra canadensis* (River otter).

**Table 11-3.** Parameters Used in the Base Patch Model (Equation 11-1) for Each Species and Life History Data Used to Calculate and Estimate Model Parameters

| Species | $b$, mass, lb (kg) | $r$, population growth rate | $K$, maximum density, number/acre (number/ha) | $D_{max}$, maximum dispersal distance, mi (km) | $f$, litters/yr | $c$, litter size | $a$, mean lifespan, yr | HR, female home range, acre (ha) | $R_0$ net reproductive rate | $a_{max}$, maximum lifespan, yr |
|---|---|---|---|---|---|---|---|---|---|---|
| *Didelphis virginiana* | 4.3780 (1.9900) | 1.386 | 0.405 (1.000) | 6.84 (11.00) | 1.0 | 16.0 | 2.00 | 125.970 (51.000) | 16.00 | 6.00 |
| *Sorex cinereus* | 0.0077 (0.0035) | 1.040 | 97.166 (240.000) | 0.17 (0.28) [a] | 1.0 | 6.0 | 1.33 | 0.099 (0.040) | 3.99 | 1.50* |
| *Blarina brevicauda* | 0.0359 (0.0163) | 2.070 | 105.263 (260.000) | 0.04 (0.06) | 2.0 | 6.0 | 0.33 | 1.457 (0.590) | 1.98 | 2.25 |
| *Cryptotis parva* | 0.0101 (0.0046) | 1.273 | 14.980 (37.000) | 0.17 (0.27) | 2.0 | 4.5 | 1.50 | 2.964 (1.200) | 6.75 | 2.75 |
| *Scalopus aquaticus* | 0.2077 (0.0944) | 0.597 | 8.097 (20.000) | 0.62 (1.00) | 1.0 | 4.0 | 3.00 | 2.495 (1.010)* | 6.00 | 6.00† |
| *Myotis lucifugus* | 0.0136 (0.0062) | 0.149 | 0.008 (0.019) [a] | 282.72 (455.00) | 1.0 | 1.0 | 12.00* | 1432.600 (580.000) | 6.00 | 30.00 |
| *Myotis sodalis* | 0.0145 (0.0066) | 0.174 | 0.037 (0.091) [b] | 30.11 (48.45) [e] | 1.0* | 1.0 | 7.96 [e] | | 3.98 | 13.00 |
| *Myotis septentrionalis* | 0.0119 (0.0054) | 0.163 | 0.093 (0.230) [c] | 36.04 (58.00) | 1.0 | 1.0 | 9.64 [e] | | 4.82 | 18.00 |
| *Lasionycteris noctivagans* | 0.0229 (0.0104) | 0.241 | 0.040 (0.099) [c] | 93.21 (150.00) ‡ | 1.0* | 1.5 | 6.68 [f] | | 5.01 | |
| *Pipistrellus subflavus* | 0.0123 (0.0056) | 0.161 | 0.046 (0.113) [c] | 80.16 (129.00) † | 1.0§ | 1.0 | 10.00 | | 5.00 | 10.00 |
| *Eptesicus fuscus* | 0.0354 (0.0161) | 0.167 | 0.058 (0.144) [c] | 60.89 (98.00) | 1.0 | 1.0 | 9.00* | | 4.50 | 19.00 |

**Table 11-3.** Parameters Used in the Base Patch Model (Equation 11-1) for Each Species and Life History Data Used to Calculate and Estimate Model Parameters (continued)

| Species | $b$, mass, lb (kg) | $r$, population growth rate | $K$, maximum density, number/acre (number/ha) | $D_{max}$, maximum dispersal distance, mi (km) | $f$, litters/yr | $c$, litter size | $a$, mean lifespan, yr | HR, female home range, acre (ha) | $R_0$, net reproductive rate | $a_{max}$, maximum lifespan, yr |
|---|---|---|---|---|---|---|---|---|---|---|
| **Lasiurus borealis** | 0.0271 (0.0123) | 0.321 | 0.028 (0.068)[a] | 298.26 (480.00) * | 1.0 | 3.0 | 7.55[e] | 414.960 (168.000) | 11.33 | 12.00[†] |
| **Lasiurus cinereus** | 0.0627 (0.0285) | 0.254 | 0.005 (0.013)[a] | 99.33 (159.85)[a] | 1.0 | 2.0* | 8.34[e] | 1976.000 (800.000) | 8.34 | 14.00[†] |
| **Nycticeius humeralis** | 0.0222 (0.0101) | 0.347 | 0.013 (0.031)[c] | 339.89 (547.00) | 1.0 | 2.0 | 2.00 | | 2.00 | 6.00 |
| **Sylvilagus floridanus** | 2.7548 (1.2522) | 1.665 | 3.036 (7.500) | 16.16 (26.00) | 3.0 | 5.4 | 1.50 | 4.199 (1.700) | 12.15 | 4.00 |
| **Lepus townsendii** | 7.8835 (3.5834) | 1.042 | 0.109 (0.270)* | 12.25 (19.71)[e] | 3.0* | 4.5* | 2.83[d] | | 19.11 | 8.00* |
| **Tamias striatus** | 0.2407 (0.1094) | 1.268 | 20.243 (50.000) | 1.35 (2.18)[a] | 2.0 | 4.0 | 1.30 | 2.470 (1.000) | 5.20 | 12.00 |
| **Marmota monax** | 7.6164 (3.4620) | 0.520 | 0.860 (2.125) | 0.19 (0.30) | 1.0 | 4.0 | 4.00 | 7.657 (3.100) | 8.00 | 6.00 |
| **Spermophilus franklinii** | 0.7641 (0.3473) | 1.071 | 30.364 (75.000) | 1.17 (1.88)[a] | 1.0 | 7.0 | 1.62[d] | 1.951 (0.790) | 5.67 | |
| **Spermophilus tridecemlineatus** | 0.3036 (0.1380) | 1.269 | 20.243 (50.000) | 0.87 (1.40)[a] | 1.0 | 8.0 | 1.30[d] | 1.235 (0.500) | 5.19 | |
| **Sciurus carolinensis** | 1.1482 (0.5219) | 0.806 | 2.024 (5.000) | 4.97 (8.00) | 2.0 | 3.0 | 2.50 | 7.657 (3.100) | 7.50 | 8.00 |
| **Sciurus niger** | 1.7670 (0.8032) | 0.973 | 2.024 (5.000) | 46.60 (75.00) | 2.0 | 3.5 | 2.00 | 8.769 (3.550) | 7.00 | 7.00 |

**Table 11-3.** Parameters Used in the Base Patch Model (Equation 11-1) for Each Species and Life History Data Used to Calculate and Estimate Model Parameters (continued)

| Species | $b$, mass, lb (kg) | $r$, population growth rate | $K$, maximum density, number/acre (number/ha) | $D_{max}$, maximum dispersal distance, mi (km) | $f$, litters/yr | $c$, litter size | $a$, mean lifespan, yr | HR, female home range, acre (ha) | $R_o$, net reproductive rate | $a_{max}$, maximum lifespan, yr |
|---|---|---|---|---|---|---|---|---|---|---|
| *Glaucomys volans* | 0.1463 (0.0665) | 0.623 | 8.502 (21.000) | 0.25 (0.40) | 2.0 | 4.5 | 5.00 | 1.136 (0.460) | 22.50 | 8.00 |
| *Geomys bursarius* | 0.5086 (0.2312) | 0.814 | 1.604 (3.962)[d] | 2.76 (4.44)[d] | 1.0* | 4.5* | 1.47[d] | | 3.30 | |
| *Perognathus flavescens* | 0.0194 (0.0088) | 1.043 | 44.991 (111.129)[d] | 0.04 (0.07)[a] | 1.5* | 4.0* | 0.67[d] | 0.012 (0.005) | 2.01 | 5.00* |
| *Castor canadensis* | 35.6400 (16.2000) | 0.358 | 8.927 (22.050) | 11.18 (18.00) | 1.0 | 3.5 | 7.00* | 502.447 (203.420)* | 12.25 | 21.00 |
| *Reithrodontomys megalotis* | 0.0238 (0.0108) | 3.248 | 2.401 (5.930)[†] | 0.82 (1.32)[a] | 7.0* | 4.0 | 0.70[d] | 1.136 (0.460)* | 9.87 | 1.25* |
| *Peromyscus maniculatus* | 0.0365 (0.0166) | 1.767 | 43.725 (108.000) | 0.31 (0.50) | 2.0 | 5.0 | 0.42 | 1.482 (0.600) | 2.10 | 2.75 |
| *Peromyscus leucopus* | 0.0477 (0.0217) | 1.102 | 19.717 (48.700) | 0.27 (0.43)[a] | 2.0 | 4.0 | 0.38 | 0.200 (0.081) | 1.52 | 1.00 |
| *Onychomys leucogaster* | 0.0528 (0.0240) | 2.666 | 2.409 (5.949)[a] | 2.32 (3.73)[a] | 6.0** | 3.8** | 0.85[d] | 5.681 (2.300)** | 9.73 | 3.00** |
| *Microtus pennsylvanicus* | 0.0854 (0.0388) | 4.170 | 56.275 (139.000) | 0.17 (0.28)[a] | 4.0 | 6.0 | 0.33 | 0.099 (0.040)[†] | 3.96 | 1.65* |
| *Microtus ochrogaster* | 0.0772 (0.0351) | 2.303 | 146.559 (362.000) | 0.20 (0.32)[a] | 5.0 | 4.0 | 1.00 | 0.124 (0.050) | 10.00 | 4.00 |
| *Microtus pinetorum* | 0.0598 (0.0272) | 0.604 | 49.798 (123.000) | 37.28 (60.00) | 3.0 | 3.0 | 0.26 | 0.247 (0.100) | 1.17 | 1.00 |

**Table 11-3.** Parameters Used in the Base Patch Model (Equation 11-1) for Each Species and Life History Data Used to Calculate and Estimate Model Parameters (continued)

| Species | b, mass, lb (kg) | r, population growth rate | K, maximum density, number/acre (number/ha) | $D_{max}$, maximum dispersal distance, mi (km) | f, litters/yr | c, litter size | a, mean lifespan, yr | HR, female home range, acre (ha) | $R_0$, net reproductive rate | $a_{max}$, maximum lifespan, yr |
|---|---|---|---|---|---|---|---|---|---|---|
| Ondatra zibethicus | 2.4354 (1.1070) | 0.870 | 35.223 (87.000) | 20.51 (33.00) | 2.0 | 6.0 | 3.50 | 7.657 (3.100) | 21.00 | 10.00 |
| Synaptomys cooperi | 0.0605 (0.0275) | 0.811 | 5.668 (14.000) | 0.07 (0.11) | 2.0 | 3.0 | 0.50* | 0.148 (0.060) | 1.50 | 1.00* |
| Rattus norvegicus | 0.5493 (0.2497) | 3.892 | 16.672 (41.180)* | 1.04 (1.67)[a] | 4.0 | 7.0 | 0.50* | 1.630 (0.660)* | 7.00 | 3.00* |
| Mus musculus | 0.0396 (0.0180) | 4.382 | 21.457 (53.000) | 0.62 (1.00) | 5.0 | 6.0 | 0.42 | 1.482 (0.600) | 6.30 | 5.00 |
| Zapus hudsonius | 0.0414 (0.0188) | 1.609 | 12.146 (30.000) | 0.31 (0.50) | 2.0 | 5.0 | 1.00 | 1.482 (0.600) | 5.00 | 5.00 |
| Canis latrans | 25.5464 (11.6120) | 0.482 | 0.001 (0.002) | 102.53 (165.00) | 1.0 | 6.0 | 6.00† | 12350.000 (5000.000) | 18.00 | 10.00 |
| Vulpes vulpes | 8.7417 (3.9735) | 0.916 | 0.004 (0.010) | 244.82 (394.00) | 1.0 | 5.0 | 1.00 | 992.940 (402.000) | 2.50 | 3.00 |
| Urocyon cinereoargenteus | 8.8202 (4.0092) | 0.693 | 0.006 (0.015)† | 51.41 (82.74)[a] | 1.0 | 4.0 | 1.00 | 708.890 (287.000) | 2.00 | 10.00 |
| Procyon lotor | 12.3614 (5.6188) | 0.486 | 0.101 (0.250) | 164.66 (265.00) | 1.0 | 3.5 | 4.00 | 27.170 (11.000) | 7.00 | 13.00 |
| Mustela erminea | 0.3784 (0.1720) | 0.896 | 0.006 (0.016) | 3.48 (5.60) | 1.0 | 6.0 | 2.00 | 28.899 (11.700) | 6.00 | |
| Mustela nivalis | 0.0704 (0.0320) | 1.272 | 1.012 (2.500) | 2.44 (3.93)[a] | 2.0 | 3.5 | 0.91[d] | 6.175 (2.500) | 3.20 | |

**Table 11-3.** Parameters Used in the Base Patch Model (Equation 11-1) for Each Species and Life History Data Used to Calculate and Estimate Model Parameters (continued)

| Species | b, mass, lb (kg) | r, population growth rate | K, maximum density, number/acre (number/ha) | $D_{max}$, maximum dispersal distance, mi (km) | f, litters/yr | c, litter size | a, mean lifespan, yr | HR, female home range, acre (ha) | $R_0$, net reproductive rate | $a_{max}$, maximum lifespan, yr |
|---|---|---|---|---|---|---|---|---|---|---|
| *Mustela frenata* | 0.2077 (0.0944) | 0.621 | 0.253 (0.625) | 3.81 (6.13) [a] | 1.0 | 6.0 | 4.00 | 12.350 (5.000) | 12.00 | 3.00* |
| *Mustela vison* | 1.2133 (0.5515) | 0.620 | 0.034 (0.085) | 19.88 (32.00) [†] | 1.0 | 5.0 | 3.50 | 34.580 (14.000) | 8.75 | 6.00* |
| *Taxidea taxus* | 15.3919 (6.9963) | 0.424 | 0.001 (0.002) | 29.83 (48.00) * | 1.0 | 3.0 | 4.50 | 585.390 (237.000) | 6.75 | 14.00 |
| *Spilogale putorius interrupta* | 0.9900 (0.4500) | 0.672 | 0.020 (0.050)[†] | 4.04 (6.50) * | 1.0* | 5.0* | 3.00* | 160.550 (65.000)* | 7.50 | 5.00* |
| *Mephitis mephitis* | 3.8117 (1.7326) | 1.099 | 0.097 (0.240) | 4.97 (8.00) | 1.0 | 6.0 | 1.00 | 40.014 (16.200) | 3.00 | 4.00 |
| *Odocoileus virginianus* | 110.0000 (50.0000) | 0.245 | 0.156 (0.385) | 34.18 (55.00) | 1.0 | 1.5 | 2.40** | 321.100 (130.000) | 1.80 | 15.00 |
| *Lynx rufus* | 19.7780 (8.9900) | 0.402 | 0.000 (0.001) | 300.12 (483.00) | 1.0 | 2.5 | 4.00 | 21365.500 (8650.000) | 5.00 | 24.00 |
| *Lutra canadensis* | 14.7400 (6.7000) | 0.288 | 0.004 (0.010) | 99.42 (160.00) | 1.0 | 2.5 | 8.00 | 3112.200 (1260.000) | 10.00 | 10.00 |

*Notes:* Female body mass (b, kilograms) is from Silva and Downing (1995). Population growth rate (r) is calculated according to $R_0 = f \cdot c_2 \cdot a$. Maximum dispersal distance ($D_{max}$, kilometers), maximum density (K, number/hectare), litter frequency (f, litters/year), litter size (c, number), mean life span (a, years), female home range size (HR, hectares), and maximum life span ($a_{max}$, years) are from Baker (1983) except where noted.

For the common name of each species, please refer to the notes in Table 11-2.

*Schwartz and Schwartz (1981)       †Jackson (1961)       ‡Hazard (1982)       §Burt (1972)       **Caire et al. (1989)

[a] Estimated from female home range size

[b] Estimated as the average among Vespertilionids

[c] Estimated from female home range size and maximum dispersal distance

[d] Estimated from female body mass

[e] Estimated from maximum life span

[f] Estimated from maximum life span and maximum dispersal distance

We also estimated population viability from relative density based on other simulations. Results from community model simulations (not shown here) using the baseline Buck Creek and Walnut Creek landscapes established a relationship between time to extinction ($t_e$, years) and relative density ($d_r$, unitless): $t_e = 3197.1 d_r^{0.15}$ ($r^2 = 0.48$, $n = 80$). The estimated time to extinction provided an easily computed measure quantifying population viability and the probability of extinction (assuming that probability of extinction is related to time to extinction via Poisson) for comparison in simulations.

## Evaluating Our Results

Landscape composition affected predicted mammal community composition in model simulations. Diversity (richness) and density differed with alternative scenarios for both watersheds (Figure 11-1).

Generally, landscape futures supporting high absolute mammal diversity also had high relative diversity (Figure 11-1a,b). As a result, futures with the most diverse mammal community were also most efficient in supporting their maximum potential community diversity. In contrast, futures with the highest absolute densities (and thus mammal abundance) had the lowest mean relative densities (Figure 11-1c, d). Grassland habitat (Categories 11, 13, and 14; Table 11-1) and woodland habitat (Categories 17, 18, 19, and 20; Table 11-1) represent the principal wild mammal habitat (Table 11-2). A strong species–area relationship existed between the predicted number of breeding species ($B$, unitless) present and grassland patch size ($A_g$, hectares) ($B = 4.96 \cdot A_g^{0.27}$, $r^2 = 0.89$) and woodland patch size ($A_w$, hectares) ($B = 3.63 \cdot A_w^{0.3}$, $r^2 = 0.84$). In all scenarios, larger grassland and woodland patches supported greater diversity (Figure 11-2). Increased species richness in larger patches, however, intensified species interactions (i.e., predation and competition), which reduced the effectiveness with which most species exploited habitat and therefore reduced densities relative to carrying capacities (i.e., relative density; Figure 11-2). Negative effects for prey and competitor species outweighed benefits of interactions in larger patches for predators. Even though scenarios that supported high diversity supported high absolute mammal densities, species interactions prevented species from reaching much higher densities. Consequently, relative densities were lower.

Low relative density can mean low viability for a population. Because time to extinction is a function of relative density, a trade-off exists between increased risk of extinction and increased species richness in large habitat patches (Figure 11-2). Low relative densities in Scenario 2 and Scenario 3 landscape futures suggest that despite the rich community, many species may still be vulnerable to extinction in these landscapes. Landscapes characterized by patches with high diversity that are relatively few and isolated may have higher extinction risks for some species (Figures 11-3 through 11-6).

## The Value of the Integrated, Landscape-Level Approach

Although the benefits of integrated, landscape-based approaches are widely recognized among conservation biologists and ecologists (Selman and Doar 1992), practical applications, models, and guidelines for land managers are lacking (Suter 1998; but see Dale and Haeuber 2001). Most efforts for developing landscape-level management strategies have focused on conservation planning and on maintaining ecological

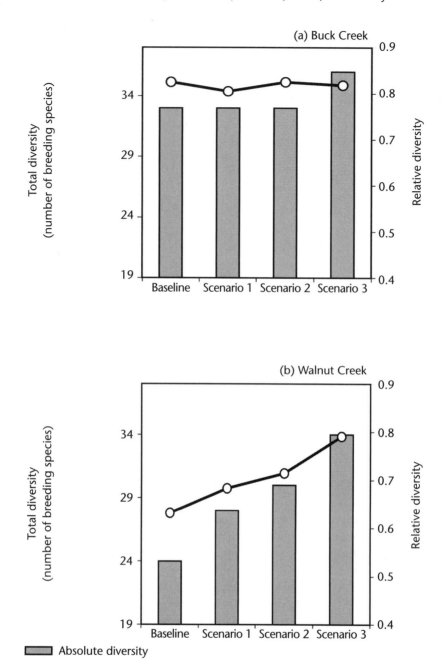

**Figure 11-1. (a,b)** Absolute and Relative Diversity (a,b) and Density (c,d) in the Study Watersheds for Each Landscape Scenario

**Color Figure 1.** This wetland in Doolittle Prairie, Story County, Iowa, exemplifies how remnant biodiversity complements water quality in the Corn Belt. Photo credit: J.A. Dowdell

**Color Figure 2.** Economist Steve Polasky and avian ecologist Kathryn Freemark Lindsay, contributors to Part 2 of this book, work together in Iowa's Buck Creek watershed to imagine how the future landscape could remain productive and profitable for agriculture and improve water quality and enhance native biodiversity at the same time. Photo credit: D. Hellekson

**Color Figure 3.** This riparian buffer along Bear Creek in Boone County, Iowa, shows how new land covers that protect water quality can also enhance biodiversity. It is part of an Iowa State University research experiment. Photo credit: J.A. Dowdell

**Color Figure 4.** An aerial view of baseline conditions in north central Buck Creek watershed, at the future site of the bioreserve proposed in Scenario 3. Under baseline conditions, farmers noticed that the existing pasture and stream banks were suffering from erosion and needed to be improved. They gave this landscape an average rating of 2.10/5.00. Photo credit: R.C. Corry.

**Color Figure 5.** Scenario 3 in Buck Creek watershed in 2025. This formerly eroding pasture dissected by the upper reaches of Buck Creek was designated a permanent bioreserve with a wide wooded riparian buffer under Scenario 3, which emphasizes enhanced biodiversity. The bioreserve is buffered by best management practices that are designed to enhance biodiversity, including strip intercropping with perennial prairie strips, which could be harvested for native plant seed, as assumed in Chapter 7 or for biofuels. In addition, the upstream bioreserve detains and infiltrates storm water, relieving downstream flooding. The landscape is highly varied in land cover; wildlife is abundant; and recreational opportunities are available on the trails and bioreserves, as described in Chapter 4. There are fewer farmers than in Scenario 2, but many of the farmsteads are inhabited by nonfarmers. The public finds this landscape particularly appealing for recreation and rural living opportunities and is willing to pay for public benefits from environmentally beneficial practices. Farmers gave this landscape a rating of 3.28/5.00. Simulation credit: R.C. Corry

**Color Figure 6.**   The baseline landscape in Walnut Creek watershed, Story County, Iowa, where highly productive soils and efficient tile drainage have allowed more than 80 percent of the watershed to be used for corn and soybean row crop agriculture. Less than 1 percent of the land was  enrolled in the CRP in 1994 (the baseline year). This landscape reflects trends since 1970—larger fields and larger, fewer farms. Conventional production technology is used, and livestock operations are moving toward concentrated animal feeding operations (CAFOs), primarily in a few areas of the state. Iowa farmers gave this an average rating of 3.22 / 5.00  to indicate their preferences for the future of the people of Iowa.   Photo credit: R.C. Corry

**Color Figure 7.**   Scenario 1 in Walnut Creek watershed in 2025 emphasizes increased commodity production and converts more land to corn and soybean fields in bigger fields. Precision agriculture and conservation tillage are pervasive. Livestock are concentrated in CAFOs in a few areas of the state, but not in this watershed. Even fewer farms and less woodland remain. Rural population losses continue at previous rates, with small towns and community institutions at risk. Average rating by Iowa farmers: 2.66/5.00.
Simulation credit: M. Sundt.

**Color Figure 8.** Scenario 2 in Walnut Creek watershed in 2025 emphasizes improved water quality and incentivizes adoption of perennial cover crops for rotational grazing, which is fenced from nearby streams. A wider stream buffer includes off-channel floodplain wetlands. Precision farming and tile drainage supplement the rotational grazing technology, but tile drainage is detained before entering streams. More farms and more woodland remain. Rural population stabilizes with some nonfarm residents and more farmers remaining. Community institutions are stable. Iowa farmers gave this image an average rating of 3.56/5.00. Simulation credit: M. Sundt.

**Color Figure 9.** Scenario 3 in Walnut Creek watershed in 2025 emphasizes enhanced biodiversity by setting aside some land in a permanent riparian woodland bioreserve and by adopting new enterprises and practices that incorporate perennial strips within row crop agriculture. It also includes organic production targeted to a biodiversity zone that connects and buffers bioreserves and stream corridors. Livestock are concentrated in CAFOs in a few areas of the state, but not in this watershed. There are fewer farms, but a larger rural population because more nonfarm families want to live in this high-amenity landscape. Small towns and community institutions are thriving. This was rated highest of all the images shown to Iowa farmers, with a mean score of 3.78/5.00. Simulation credit: M. Sundt.

**Color Figure 10.** An aerial view of the Buck Creek watershed baseline landscape. This rolling, more erodible watershed had more land in woodland, pasture or hay, and more land enrolled in the Conservation Reserve Program (CRP) in 1994—about 16 percent. Conventional production technology is used. Livestock operations are moving toward CAFOs, primarily in areas of the state outside Buck Creek watershed, where grazing continues on the rolling terrain. Noting its susceptibility to erosion, Iowa farmers gave this baseline landscape an average rating of 2.53/5.00.   Photo credit: R.C. Corry

**Color Figure 11.** Scenario 1 in Buck Creek watershed in 2025 emphasizes increased commodity production and converts more highly productive land from perennial cover, including the CRP, to corn and soybean fields. Even fewer farms and much less woodland remain. Precision agriculture and conservation tillage are used pervasively, but farmers voiced their concerns about erosion and water quality, especially in comparison with the flat landscape of the Walnut Creek watershed. Livestock are concentrated in CAFOs in a few areas of the state, but there are none in this watershed. Rural population losses continue at previous rates, and small towns and community institutions are disappearing. In this scenario, the public sees the landscape as boring, but perceives this form of agriculture as safe. The scenario assumes that the public is not particularly concerned about environmental effects, and continues to be willing to pay commodity subsidies. Iowa farmers gave this landscape their lowest average rating:1.47/5.00. Simulation credit: R.C. Corry.

**Color Figure 12.** Scenario 2 in Buck Creek watershed in 2025 emphasizes improved water quality and incentivizes more wide-spread adoption of perennial cover crops for rotational grazing, which is fenced from nearby streams. Woodlands are maintained for grazing and livestock shade as well. Steeper working lands near streams are in pasture or hay and a wider stream buffer includes off-channel floodplain wetlands. Grazing livestock enterprises lead to the need for more farms and farm buildings than in Scenarios 1 and 3. The public finds the variety of land cover and grazing livestock appealing, and the landscape is attractive for recreation and tourism. The public sees environmental benefits from the mixed grain and livestock enterprises and is willing to pay for subsidies. Simulation credit: R.C. Corry.

**Color Figure 13.**   Scenario 1 in west central Walnut Creek watershed, where the tile-drained, highly productive soils are well suited to corn and soybean production. Because the number of farms decreases by 50 percent in Scenario 1 compared with the base-line, there is no visible farmstead in this landscape. This simulation was not included in the Iowa farmer interviews (Chapter 6). Simulation credit: R.C. Corry

**Color Figure 14.**   Scenario 2 in west central Walnut Creek watershed leads to adoption of  some mixed-grain livestock farming with rotational grazing. Farmers are assumed to employ perennial cover as pasture or hay near streams to meet water quality stan-dards. Compared with Scenarios 1 and 3, more  farmers and farmsteads are needed to manage the rotational grazing enterprises, and a farmstead is seen in the distance in this simulation. This simulation was not included in the Iowa farmer interviews (Chapter 6). Simulation credit: R.C. Corry.

**Color Figure 15.** Scenario 3 bioreserve in west central Walnut Creek watershed, where wetland soils led to this bioreserve being designed to include several types of wetland habitats. Because this bioreserve is located on the watershed boundary (see Color Figure 43), it may enhance habitat connectivity and patch size if it is adjacent to a bioreserve in the adjacent watershed. This simulation was not included in the Iowa farmer interviews (Chapter 6). Simulation credit: R.C. Corry.

**Color Figure 16.** This cornfield (under Scenario 1 in Buck Creek watershed) on productive soils with more rolling terrain shows how the emphasis on commodity production in Scenario 1 extends cultivation to lands that might have been enrolled in the CRP or in pasture under baseline conditions. Noting the healthy appearance of the corn crop, farmers rated this landscape 3.06/5.00.
Simulation credit: R.C. Corry.

**Color Figure 17.** Under the land allocation model for Scenario 2 in Buck Creek watershed, the same field is used for pasture in rotational grazing (described in Chapter 4). Pasture and hay as perennial cover contribute to improved water quality (Chapter 7), and the public is likely to find this landscape attractive. This simulation was not included in the Iowa farmer interviews (Chapter 6).
Simulation credit: R.C. Corry

**Color Figure 18.** Scenario 3 in Buck Creek watershed brings perennial strip intercropping with prairie strips to this field. This innova-tive practice was developed for Scenario 3 based on the Iowa State University field experiments in strip intercropping (described in Chapter 4). Although farmers were concerned that managing the strip pattern could be time-consuming, they consistently noted that the strip pattern with prairie grasses suggested good stewardship, and they gave this landscape high ratings—an average of 3.75/5.00. Simulation credit: R.C. Corry

**Color Figure 19.**   This field demonstrates strip intercropping of corn, soybeans, and oats as it has been implemented by Iowa farmers in experiments with Iowa State University. Seeing good stewardship here, farmers rated this landscape an average of 3.75/5.00, tying with the average rating for Color Figure 18, which also showed strip intercropping.   Photo credit: R. Cruse

**Color Figure 20.**   Prairie remnants were recognized as a native ecosystem and seen as attractive by most of the farmers interviewed. Farmers often spoke about the need to maintain some land for habitat or for biodiversity. They rated this landscape highly: 3.66/5.00.

**Color Figure 21.** Many farmsteads and a variety of land cover types currently characterizes the upper reaches of Buck Creek watershed, shown in this photograph. Appreciating the number of farmsteads and the diversity of land covers, farmers rated this landscape 3.62/5.00. Photo credit: R.C. Corry

**Color Figure 22.** This riparian corridor restoration along Bear Creek, Story County, Iowa, is part of a cooperative field experiment conducted by Iowa State University. Farmers recognized its water quality and habitat values, but some were unenthusiastic about managing land they perceived as floodplain. Its average rating was 3.47/5.00. Photo credit: T. Isenhart

**Color Figure 23.** This pasture in Buck Creek watershed was viewed critically by farmers, who noticed where it showed signs of erosion, overgrazing, or lack of care. Farmers rated this landscape 3.16/5.00. Photo credit: J.I. Nassauer

**Color Figure 24.** Rolling fields of row crops in Buck Creek watershed led farmers to comment on tillage and planting practices for conservation. Some farmers thought this land was being managed properly, but a few thought more appropriate conservation practices could have been used. On average, farmers rated this landscape 3.12/5.00. Photo credit: R.C. Corry

**Color Figure 25.** Many farmers viewed this small wetland surrounded by cultivated land critically for its poor weed management. Farmers rated this landscape 2.66/5.00. Photo credit: USDA NRCS

**Color Figure 26.** Farmers had mixed feelings about this CAFO, wondering whether there was enough land nearby to spread manure without causing pollution and mentioning their concerns about offensive odors. At the same time, they noted the financial pressures and rewards that might lead to adoption of CAFO technology. On average, however, the farmers rated this landscape as one of the least desirable for the future of the people of Iowa at 1.84/5.00. Photo credit: J.I. Nassauer

**Color Figure 27.** In this photograph, farmers noticed residential development surrounded by farmland. They also viewed the uneven color and texture in the row crops as indicating poor management or poor soils for production. Most farmers saw both as undesirable, and this landscape had the lowest average rating: 1.84/5.00. Photo credit: J.I. Nassauer

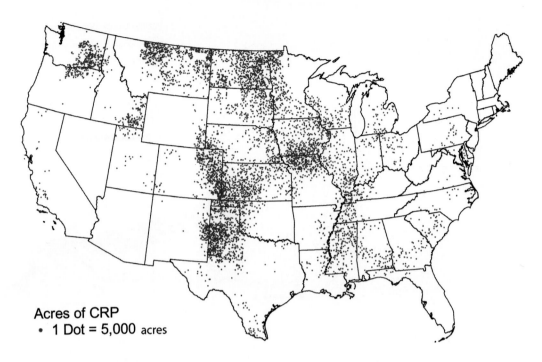

Acres of CRP
• 1 Dot = 5,000 acres

**Color Figure 28.** Geographic distribution of the CRP is shown in this map and discussed in Chapter 3.

*Source*: Maps created by the U.S. Department of Agriculture (USDA) Economic Research Service (ERS)

Watersheds [1]

2004 [2]

2005

2006

1. Watersheds in AK, HI, PR, Virgin Islands, Samoa, and Pacific Basin are not shown.
For more details see http://www.nrcs.usda.gov/programs/csp
2. All 2004 eligible watersheds were eligible in 2005.

**Color Figure 29.** Color Figure 29. Watersheds eligible for the Conservation Security Program (CSP) in 2006 are shown in this map and discussed in Chapter 3.

*Source*: Maps created by the USDA ERS

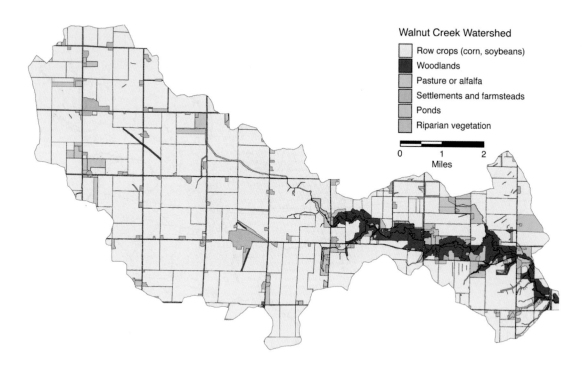

**Color Figure 30.** This map shows the baseline (1994) land cover of Walnut Creek watershed, one of two Iowa study watersheds for the Corn Belt futures project (Chapter 4). Dark green indicates woodlands, which occur almost exclusively along Walnut Creek.

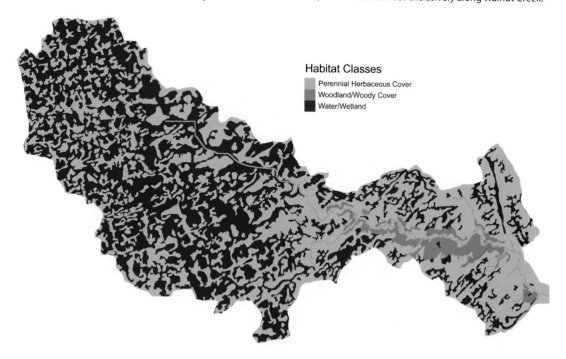

**Color Figure 31.** Past habitat of Walnut Creek watershed as inferred from soil characteristics

*Note:* Dark blue indicates the pervasiveness of wetland ecosystems in the watershed before tile drainage for agriculture.

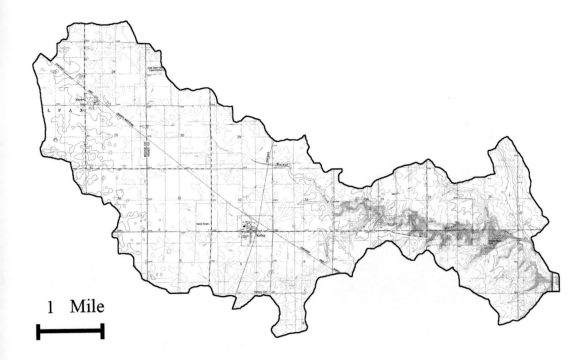

1   Mile

**Color Figure 32.**   Low relief of Walnut Creek watershed topography

*Source*: U.S. Geological Survey (USGS)

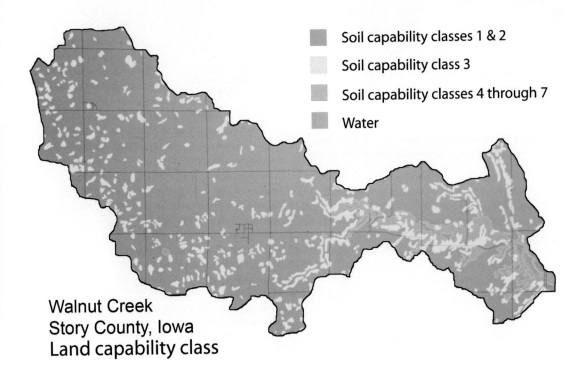

Soil capability classes 1 & 2

Soil capability class 3

Soil capability classes 4 through 7

Water

Walnut Creek
Story County, Iowa
Land capability class

**Color Figure 33.**   High productivity of Walnut Creek watershed soils (see Chapter 4)

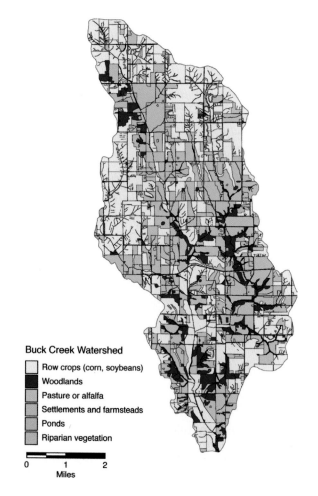

**Buck Creek Watershed**

- Row crops (corn, soybeans)
- Woodlands
- Pasture or alfalfa
- Settlements and farmsteads
- Ponds
- Riparian vegetation

0    1    2
Miles

**Color Figure 34.** Baseline (1994) land cover of Buck Creek watershed; pervasive perennial cover as pasture or CRP on steeper slopes

Habitat Classes
- Perennial Herbaceous Cover
- Woodland/Woody Cover
- Water/Wetland

**Color Figure 35.** Past habitat of Buck Creek watershed shows prairie uplands with pervasive woodlands along lower stream reaches and steeper slopes. Wetland habitats were relatively rare in Buck Creek watershed compared with Walnut Creek.

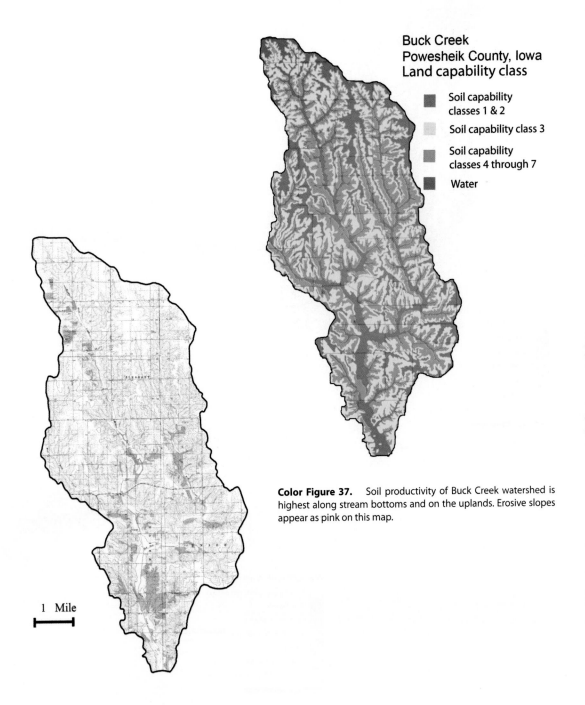

**Buck Creek
Powesheik County, Iowa
Land capability class**

■ Soil capability classes 1 & 2

▨ Soil capability class 3

■ Soil capability classes 4 through 7

■ Water

**Color Figure 37.**  Soil productivity of Buck Creek watershed is highest along stream bottoms and on the uplands. Erosive slopes appear as pink on this map.

1 Mile

**Color Figure 36.**  High relief of Buck Creek watershed topography compared with Walnut Creek watershed

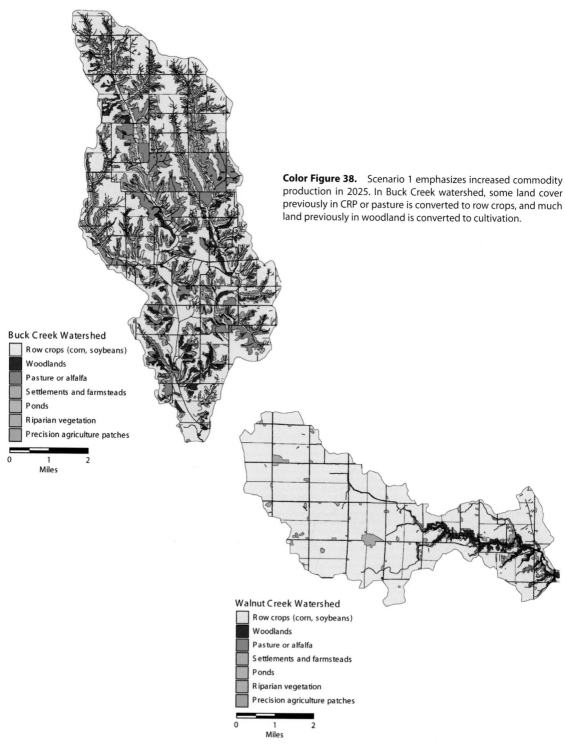

**Buck Creek Watershed**

- Row crops (corn, soybeans)
- Woodlands
- Pasture or alfalfa
- Settlements and farmsteads
- Ponds
- Riparian vegetation
- Precision agriculture patches

0    1    2
Miles

**Color Figure 38.** Scenario 1 emphasizes increased commodity production in 2025. In Buck Creek watershed, some land cover previously in CRP or pasture is converted to row crops, and much land previously in woodland is converted to cultivation.

**Walnut Creek Watershed**

- Row crops (corn, soybeans)
- Woodlands
- Pasture or alfalfa
- Settlements and farmsteads
- Ponds
- Riparian vegetation
- Precision agriculture patches

0    1    2
Miles

**Color Figure 39.** Scenario 1 emphasizes increased commodity production in 2025. In Walnut Creek watershed, land cover is similar to the baseline condition (Color Figure 30), but woodlands on highly productive soils have been converted to row crops. Precision agriculture also creates some less productive patches not in row crops.

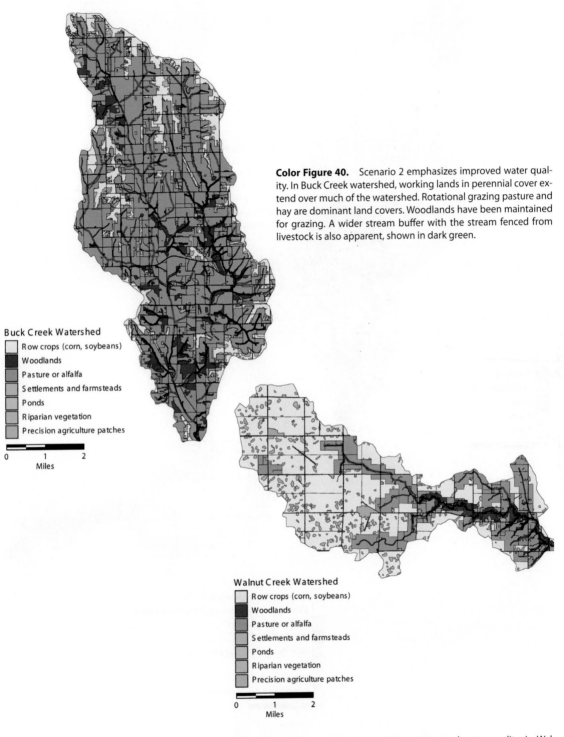

**Color Figure 40.** Scenario 2 emphasizes improved water quality. In Buck Creek watershed, working lands in perennial cover extend over much of the watershed. Rotational grazing pasture and hay are dominant land covers. Woodlands have been maintained for grazing. A wider stream buffer with the stream fenced from livestock is also apparent, shown in dark green.

Buck Creek Watershed

- Row crops (corn, soybeans)
- Woodlands
- Pasture or alfalfa
- Settlements and farmsteads
- Ponds
- Riparian vegetation
- Precision agriculture patches

0     1     2
Miles

Walnut Creek Watershed

- Row crops (corn, soybeans)
- Woodlands
- Pasture or alfalfa
- Settlements and farmsteads
- Ponds
- Riparian vegetation
- Precision agriculture patches

0     1     2
Miles

**Color Figure 41.** Scenario 2 emphasizes improved water quality. In Walnut Creek watershed, working lands in perennial cover are located along the stream, where some rotational grazing pasture and hay occur, and where woodlands remain for grazing. A wider stream buffer with the stream fenced from livestock is also apparent, shown in dark green. Perennial cover patches identified by precision agriculture dot large productive corn–soybean fields.

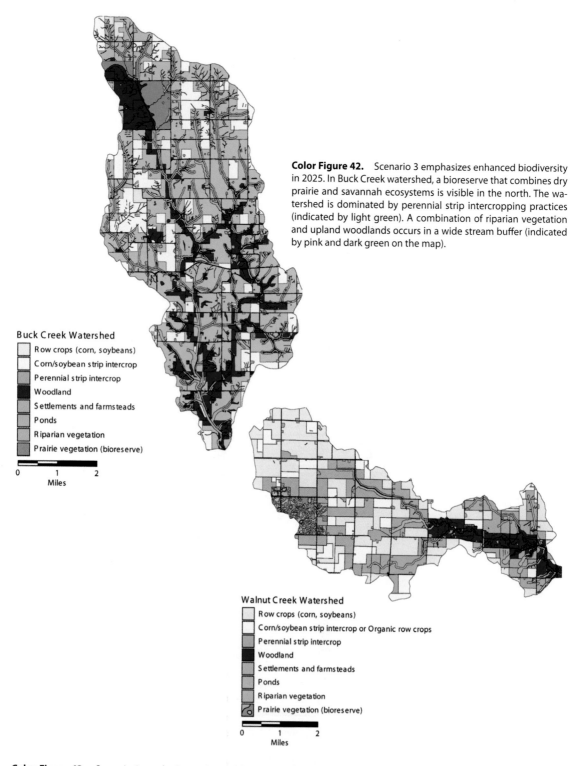

**Color Figure 42.** Scenario 3 emphasizes enhanced biodiversity in 2025. In Buck Creek watershed, a bioreserve that combines dry prairie and savannah ecosystems is visible in the north. The watershed is dominated by perennial strip intercropping practices (indicated by light green). A combination of riparian vegetation and upland woodlands occurs in a wide stream buffer (indicated by pink and dark green on the map).

**Buck Creek Watershed**
- Row crops (corn, soybeans)
- Corn/soybean strip intercrop
- Perennial strip intercrop
- Woodland
- Settlements and farmsteads
- Ponds
- Riparian vegetation
- Prairie vegetation (bioreserve)

0    1    2
Miles

**Walnut Creek Watershed**
- Row crops (corn, soybeans)
- Corn/soybean strip intercrop or Organic row crops
- Perennial strip intercrop
- Woodland
- Settlements and farmsteads
- Ponds
- Riparian vegetation
- Prairie vegetation (bioreserve)

0    1    2
Miles

**Color Figure 43.** Scenario 3 emphasizes enhanced biodiversity in 2025. In Walnut Creek watershed, two bioreserves are established—one that represents wetland ecosystems in the west, and another along the stream corridor that represents riparian ecosystems. Perennial strip intercropping practices (indicated by light green) and organic crops (light yellow) occur in a biodiversity target zone that links the stream corridor and wetland reserve in this watershed. Because of its productive soils, much of this watershed remains in a corn–soybean rotation under the land allocation model for Scenario 3 (Chapter 4).

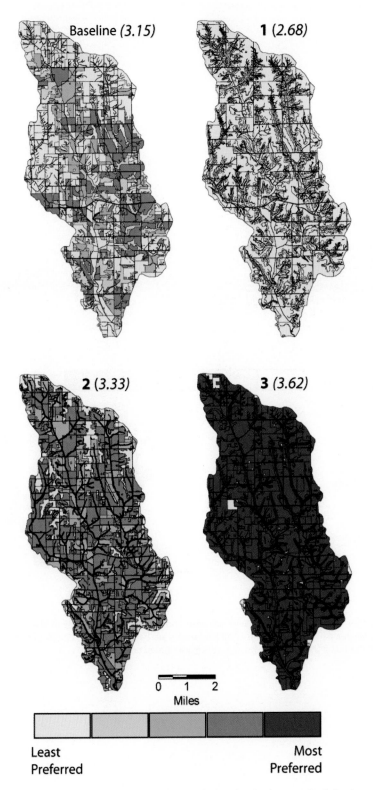

**Color Figure 44.** Farmer ratings of three alternative scenarios and the baseline landscape in Buck Creek watershed are described in Chapter 6. Mean area-weighted ratings of all land covers are based on farmer ratings of images. (Bold type indicates scenario number or name; numbers in parentheses indicate area-weighted ratings).

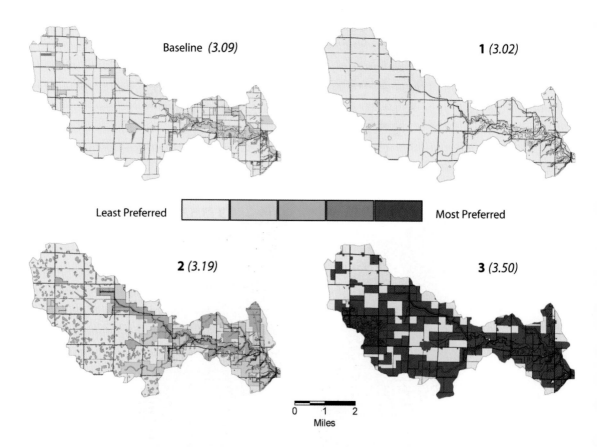

**Baseline** *(3.09)*

**1** *(3.02)*

Least Preferred                                Most Preferred

**2** *(3.19)*

**3** *(3.50)*

0    1    2
Miles

**Color Figure 45.** Farmer ratings of three alternative scenarios and the baseline landscape in Walnut Creek watershed are described in Chapter 6. Mean area-weighted ratings of all land covers are based on farmer ratings of images. (Bold type indicates scenario number or name; numbers in parentheses indicate area-weighted ratings).

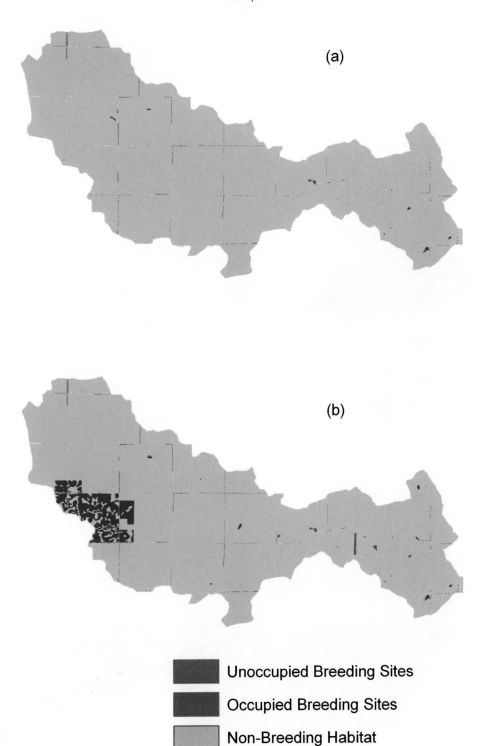

**Color Figure 46.** In the Walnut Creek watershed, breeding site occupancy for the tiger salamander in (a) the baseline and (b) Scenario 3. Red indicates sites that were unoccupied and blue indicates sites that were occupied during model years 50 to 100. In the baseline, ditches comprised the majority of habitat for this species, and most are unoccupied. In Scenario 3, most of the habitat occurred in wetlands within the prairie/wetland bioreserve and was occupied.

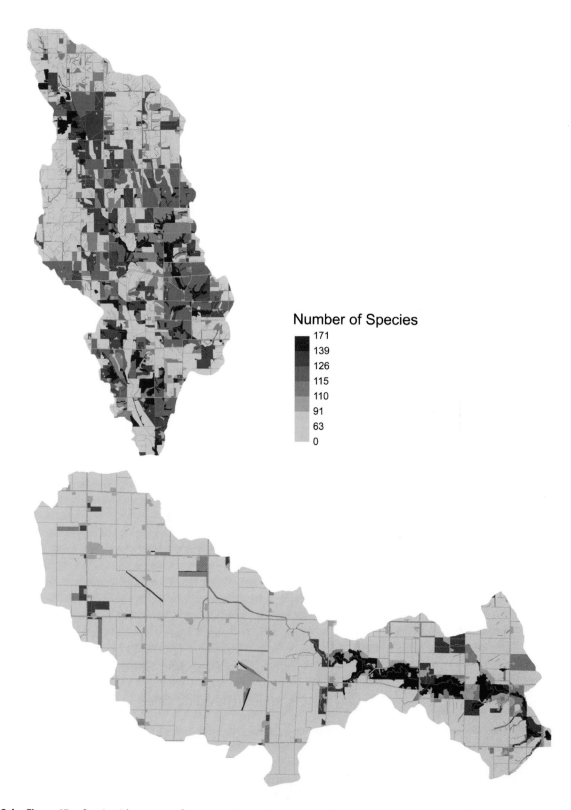

**Number of Species**

171
139
126
115
110
91
63
0

**Color Figure 47.** Species richness map of native vertebrates for the existing landscape of both watersheds.

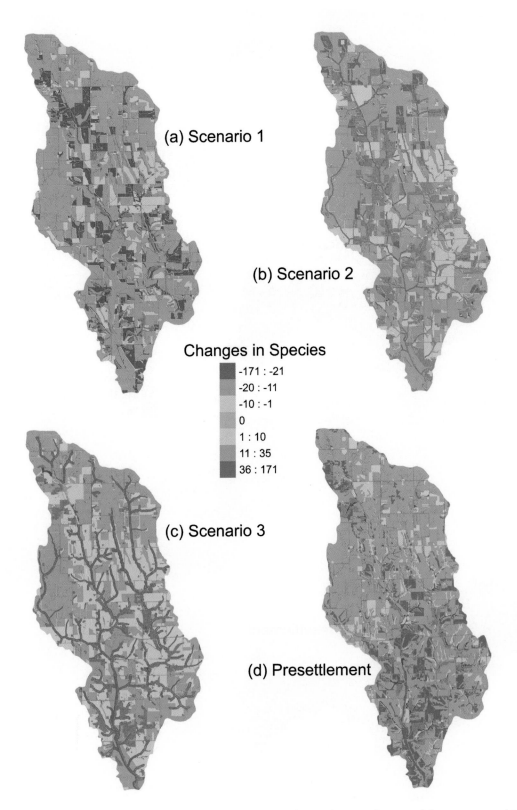

**Color Figure 48.** Species richness change maps of native vertebrates for the alternative futures and the past compared to the baseline for Buck Creek watershed, described in Chapter 13.

Color plate 49

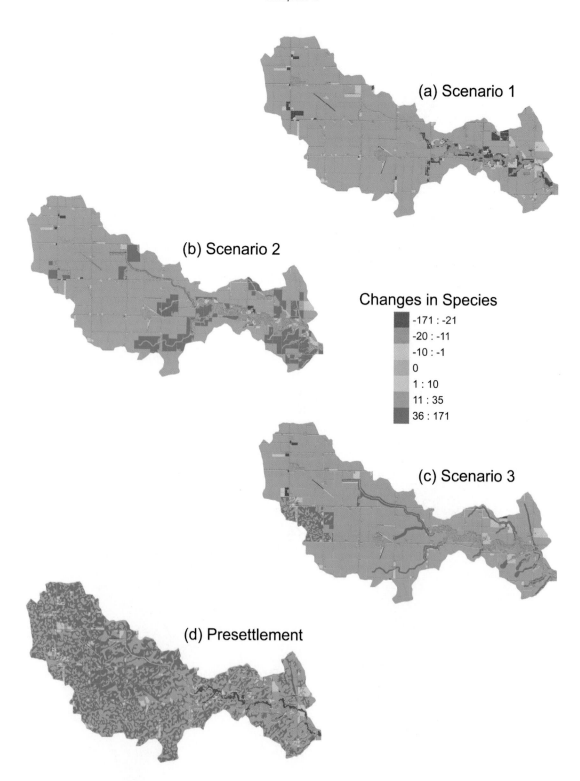

**Color Figure 49.** Species richness change maps of native vertebrates for the alternative futures and the past compared to the baseline for Walnut Creek watershed, described in Chapter 13.

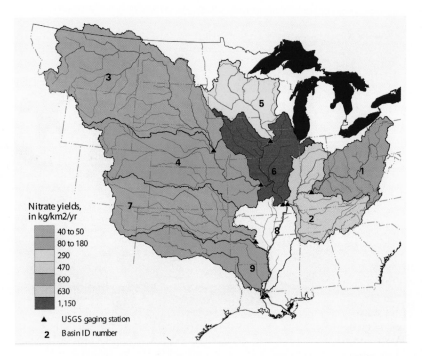

**Color Figure 50.** Nitrate yields from the sub-basins of the Mississippi River Basin are shown here and described in Chapter 15.

*Notes*: 1 = Upper Ohio, 2 = Lower Ohio, 3 = Upper Missouri, 4 = Lower Missouri, 5 = Upper Mississippi, 6 = Middle Mississippi, 7 = Arkansas, 8 = Lower Mississippi, 9 = Red and Ouachita Basins are identified by bold numbers. *Source*: Modified from Goolsby et al. (1999)

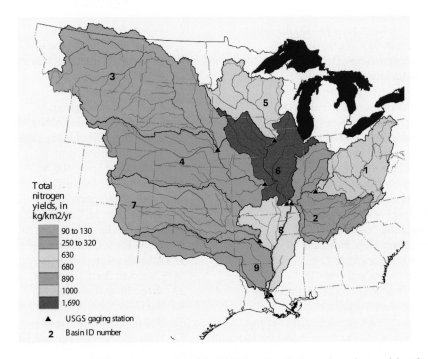

**Color Figure 51.** Total nitrogen yields from the sub-basins of the Mississippi River Basin are shown here and described in Chapter 15.

*Notes*: 1 = Upper Ohio, 2 = Lower Ohio, 3 = Upper Missouri, 4 = Lower Missouri, 5 = Upper Mississippi, 6 = Middle Mississippi, 7 = Arkansas, 8 = Lower Mississippi, 9 = Red and Ouachita Basins are identified by bold numbers. *Source*: Goolsby et al. (1999)

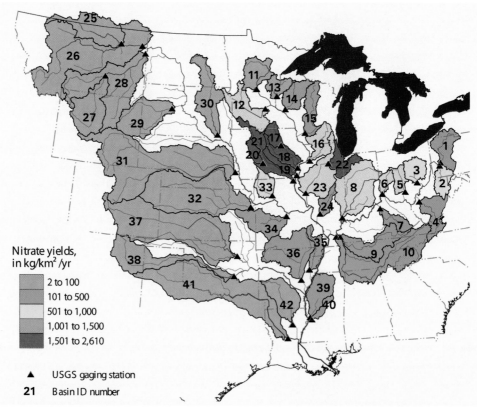

Nitrate yields,
in kg/km$^2$/yr

| | 2 to 100 |
| | 101 to 500 |
| | 501 to 1,000 |
| | 1,001 to 1,500 |
| | 1,501 to 2,610 |

▲   USGS gaging station

**21**   Basin ID number

Small Basin Sites

1 - Allegheny River, New Kensington, PA
2 - Monongahela River, Braddock, PA
3 - Muskingham River, McConnelsville, OH
4 - Kanawha River, Winfield, WV
5 - Scioto River, Higby, OH
6 - Great Miami, New Baltimore, OH
7 - Kentucky River, Lockport, KY
8 - Wabash River, New Harmony, IN
9 - Cumberland River, Grand Rivers, KY
10 - Tennessee River, Paducah, KY
11 - Mississippi River, Royalton, MN
12 - Minnesota River, Jordan, MN
13 - St. Croix River, St. Croix Falls, WI
14 - Chippewa River, Durand, WI
15 - Wisconsin River, Muscoda, WI
16 - Rock River, Joslin, IL
17 - Cedar River, Cedar Falls, IA
18 - Iowa River, Wapello, IA (includes 17)
19 - Skunk River, Augusta, IA
20 - Raccoon River, VanMeter/Des Moines, IA
21 - Des Moines, St. Francisville, MO (includes20)

22 - Illinois, River, Marseilles, IL
23 - Lower Illinois River, Valley City, IL
24 - Kaskaskia River, Venedy Station, IL
25 - Milk River, Nashua, MT
26 - Missouri River, Culbertson, MT
27 - Bighorn River, Bighorn MT
28 - Yellowstone River, Sydney, MT
29 - Cheyenne River, Cherry Creek, SD
30 - James River, Scotland, SD
31 - Platte River, Louisville, NE
32 - Kansas River, Desoto, KS
33 - Grand River, Sumner, MO
34 - Osage River, St. Thomas., MO
35 - St. Francis Bay, Riverfront, AR
36 - White River, Clarendon, AR
37 - Arkansas River, Tulsa, OK
38 - Canadian River, Calvin, OK
39 - Yazoo River, Redwood, MS
40 - Big Black River, Bovina, MS
41 - Red River, Alexandria, LA
42 - Ouachita River, Columbia, LA

**Color Figure 52.**   Spatial distribution of nitrate yields is highly varied among smaller watersheds, as described in Chapter 15. Among the highest yields are from the Skunk River watershed in Iowa, the location of the two Corn Belt futures study watersheds described in Part 2.

*Note:* Basins are identified by bold numbers.   *Source:* Modified from Goolsby et al. (1999)

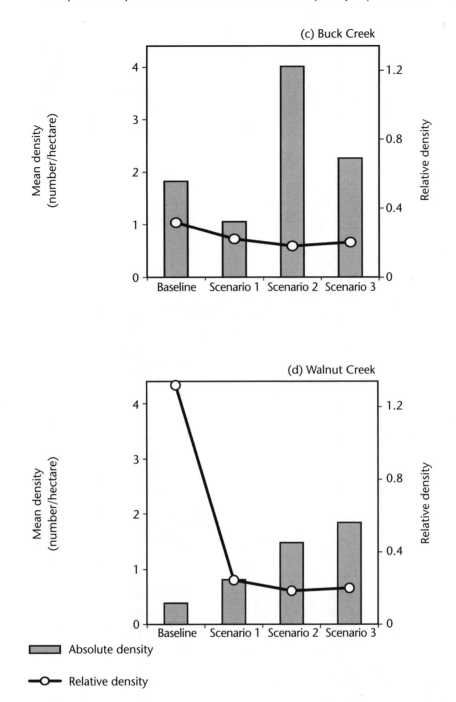

**Figure 11-1. (c,d)**    Absolute and Relative Diversity (a,b) and Density (c,d) in the Study Watersheds for Each Landscape Scenario

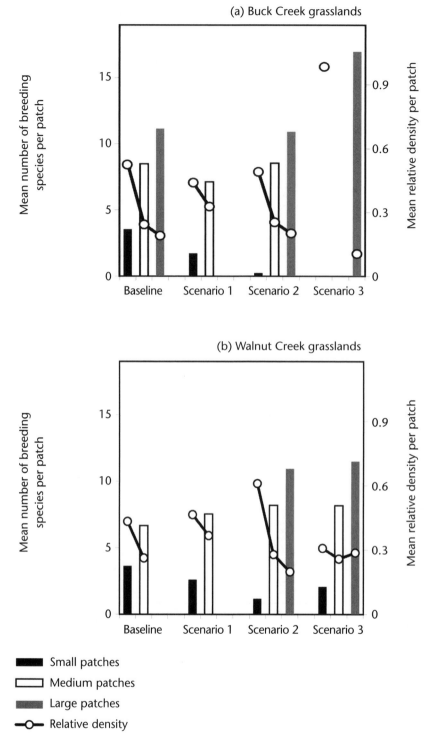

Small patches

Medium patches

Large patches

—O— Relative density

**Figure 11-2. (a,b)**   Species Diversity and Relative Density in Small, Medium, and Large Grassland (a and b) and Woodland (c and d) Patches for Each Landscape Scenario

*Notes:* Small = <1 ha; medium = between 1 and 10 ha; large = >10 ha

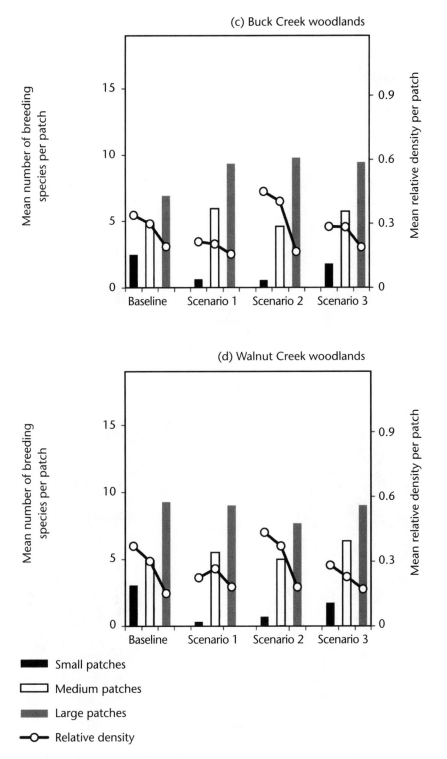

**Figure 11-2. (c,d)**    Species Diversity and Relative Density in Small, Medium, and Large Grassland (a and b) and Woodland (c and d) Patches for Each Landscape Scenario

*Notes:* Small = <1 ha; medium = between 1 and 10 ha; large = >10 ha

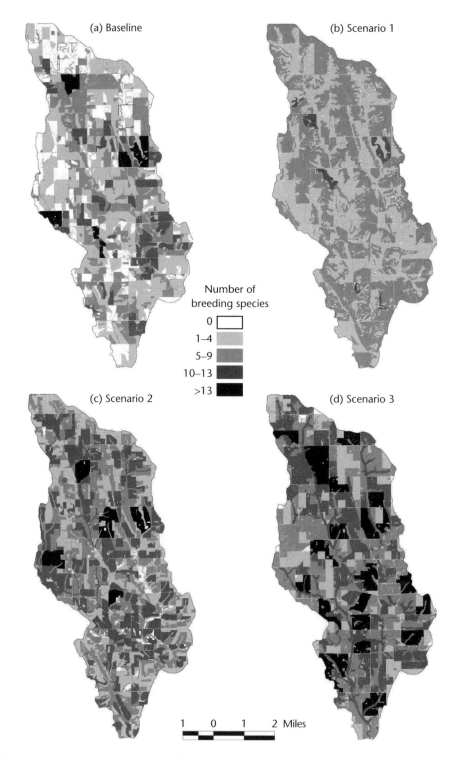

**Figure 11-3.** Contrasting Maps of Species Diversity in the Buck Creek Watershed

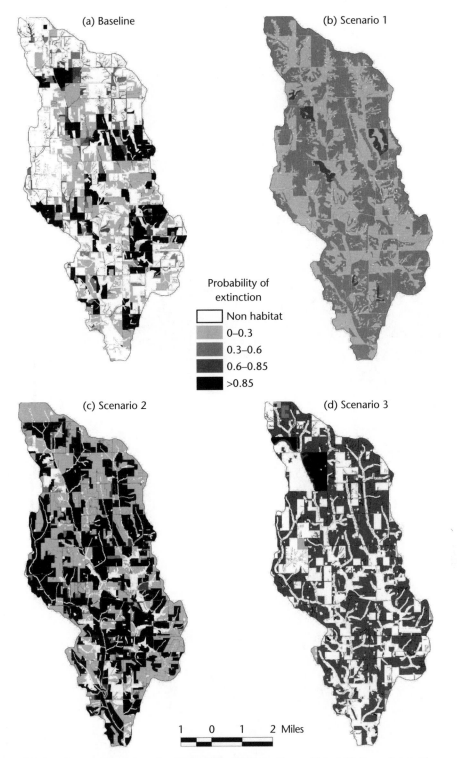

**Figure 11-4.** Contrasting Maps for Probability of Extinction within 100 Years for *Reithrodontomys megalotis* in Buck Creek Watershed

Notes: Computed from time to extinction, $_{te}$, as $P_e = e^{-100}/_{te}$ ) in (a), the baseline landscape; (b), Scenario 1; (c), Scenario 2; and (d), Scenario 3. Land cover associated with areas of high diversity in the landscape may overlap areas with high risk of extinction for some species, indicating potential conflicts for land use planning.

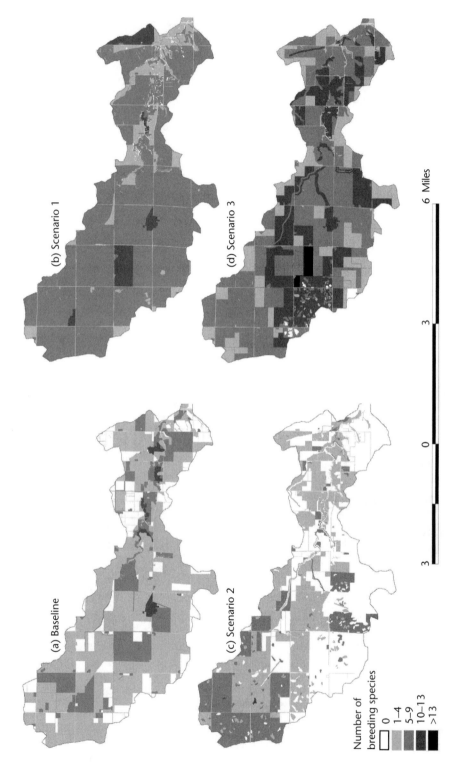

**Figure 11-5.** Contrasting Landscape Maps of Species Diversity in Walnut Creek Watershed

(b) Scenario 1

(d) Scenario 3

(a) Baseline

(c) Scenario 2

Number of
breeding species
☐ 0
1–4
5–9
10–13
>13

3    0    3    6  Miles

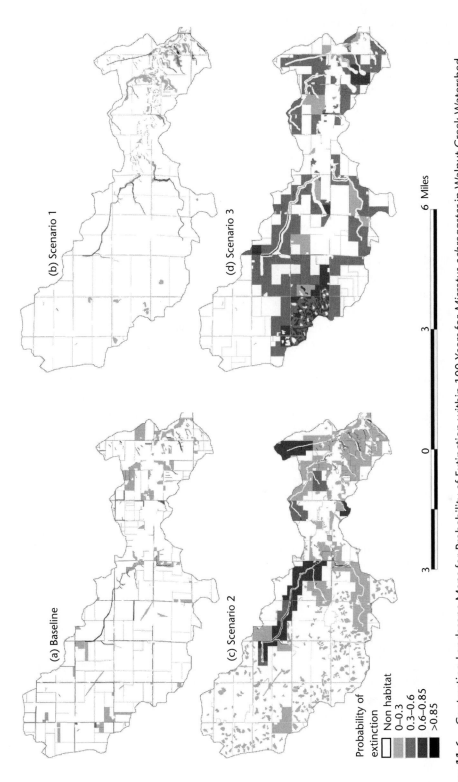

**Figure 11-6.** Contrasting Landscape Maps for Probability of Extinction within 100 Years for *Microtus ochrogaster* in Walnut Creek Watershed

*Notes:* Computed from time to extinction, $t_e$, as $P_e = e^{-100/t_e}$ in (a), the baseline landscape; (b), Scenario 1; (c), Scenario 2; and (d) Scenario 3. Land cover associated with areas of high diversity in the landscape may overlap areas with high risk of extinction for some species, indicating potential conflicts for land use planning.

diversity through general theoretical principles of habitat fragmentation (Collinge 1996); intensive single-species case studies (Rabb and Sullivan 1995); or static, habitat-based analyses (Pressey et al. 1993; White et al. 1997). The approach presented here for evaluating mammal communities in alternative landscape futures is a widely applicable methodology.

Five indices yielded systematic, quantitative measures for comparing within and among alternative landscape futures and their associated mammal communities. The indices focus on three aspects critical to managing wildlife at landscape levels: species diversity, population size, and population viability (Pressey et al. 1993). Absolute and relative diversity (Figure 11-1a,b) permitted assessment of total species diversity supported by a particular landscape as well as the extent to which potential diversity was realized in the landscape. Likewise, absolute and relative densities (Figure 11-1c,d) yielded not only basic population size comparisons, but also assessments of the efficiency with which species exploit habitat within a landscape. Time to extinction, an additional estimate from relative density, provided a measure for population viability and community stability.

Estimated absolute and relative diversity are most appropriate when management objectives emphasize maximizing species diversity. These estimates inform a more thorough evaluation of community diversity in the landscape than a traditional approach that uses only species richness. For simulated mammal communities in the alternative futures for Buck Creek, estimates of relative diversities indicated that approximately 80 percent of the potential diversity was realized in all scenarios (Figure 11-1a). Absolute diversity was highest, however, under Scenario 3. In alternative futures for Walnut Creek, not only was absolute species diversity greater in Scenarios 2 and 3 than in the Baseline and Scenario 1, but Scenarios 2 and 3 were also better at achieving diversity because relative diversity was higher (Figure 11-1b). To enhance species diversity, Scenario 3 is preferable to Scenario 2 based on estimated relative diversity.

Estimates of absolute and relative density are most helpful in cases in which population size is paramount. Absolute density is especially useful for comparing the response of a single species between and within landscapes. For instance, white-tailed deer (*Odocoileus virginianus*) densities in Buck Creek watershed were more than twice as high in Scenario 3 than in all other scenarios, and their densities were greatest in riparian woodlands. To enhance white-tailed deer populations, land management could reflect land cover patterns in Scenario 3 (Figure 11-1c), with riparian woodlands being more critical than larger woodland blocks.

Relative density can be compared both between and within landscapes. Lower mean relative densities reveal that, on average, species are less efficient at exploiting Scenario 2 and 3 landscapes as habitat than in the (lower diversity) Baseline and Scenario 1 landscapes (Figure 11-1c,d).

Time to extinction, estimated from relative density, gives us a final metric for evaluating communities at landscape scales. Time to extinction helps to identify potential high-extinction-risk areas within the landscape (Figures 11-4 and 11-6), a central focus of conservation management for rare, threatened, or endangered species (Mann and Plummer 1995).

Taken together, these five indices permit a richer comparison of community dynamics within complex landscapes. Habitat, life history traits, and species interactions can all influence community composition and dynamics. Simulations of mammal communities in alternative future landscapes under the three scenarios showed strong species–area relationships. Larger grassland and woodland habitats consistently supported higher mammal diversity (Figure 11-2). But competition and predation intensified in larger communities associated with larger habitat patches, and most species were less successful in these large patches, where densities were lower relative to carrying capacities than in smaller patches.

The trade-off between diversity and relative density suggests that patterns of community stability (Figure 11-7) may be analogous to dynamic equilibria for diversity and disturbance (Connell 1978; Huston 1979). This can be especially important for land managers balancing multiple ecological goals because habitat within a landscape that supports high diversity may overlap with zones of high probability of extinction for many species. A highly diverse and stable community in the existing landscape may depend on the presence of nearby source populations, or such a community might even be at odds with a landscape that is managed with an objective of establishing high densities and low extinction risk for one or a few particular species of concern using a single large habitat reserve. Our results indicate that even though isolated large reserves may initially exhibit high species diversities, populations in these large reserves may have lower relative densities and be more susceptible to local extinction than populations in smaller patches. Multiple endpoints for diversity and density were critical in our evaluation of the alternative scenarios for the Iowa

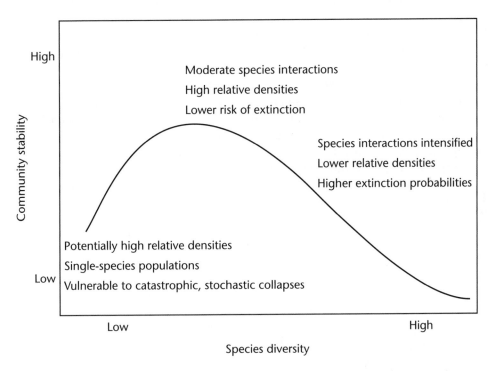

**Figure 11-7.** Hypothesized Relationship between Mammal Diversity and Community Stability*

*Based on the tradeoff between diversity and relative density (and time to extinction). Intensified competition and predation in diverse communities reduces the efficiency with which mammals exploit habitat and decrease time to extinction and population viability. At intermediate levels of diversity, risk of extinction is lower and species turnover in the community is reduced. As diversity decreases to areas with single species present, abiotic, stochastic, or catastrophic events can eliminate the population and therefore the entire community.

watersheds because landscape change can intensify or ameliorate species interactions, even qualitatively alter the very nature of the interaction (e.g., Danielson and Pulliam 1991; Danielson 1992).

## Implications for Agricultural Policy

Even though progress has been made in applying principles of landscape ecology and population ecology to land management and conservation, these principles can be even more useful (Selman and Doar 1992; Collinge 1996; Suter 1998). Land use and policy planning could deliberately utilize alternative futures modeling with the best available ecological knowledge to anticipate effects on ecological communities. Although modeling single species of special concern is easier, it is clear that indirect effects among species are common and can produce dynamics that are counterintuitive. The community model described here may advance development of more complex community-level models that include even wildlife species that may not be of particular concern to conservationists, policymakers, or the public. For example, small mammals that lack charismatic value may be extremely beneficial in managing crop pests, both insects (e.g., Clark and Young 1986) and weeds (e.g., Westerman 2003a, 2003b). How these functions are affected by landscape properties is not well known but almost certainly important (Liebman 2001).

Ecosystem management relies on the ability to evaluate community dynamics at large spatial scales, at extents at least as large as the Iowa study watersheds, as well as to think of large temporal and ecological scales (Rosenzweig 1999). Such evaluations can lead to informed decisions that are more than a short-term solution to an immediate crisis.

The approach presented here, in which indices for diversity, density, and viability were computed from population simulations that are integrated with geographic information system coverages representing landscapes, demonstrates a means of analyzing community responses in alternative landscape futures. The indices reduce and normalize population and community information for tractable, quantitative comparisons between landscapes, within landscapes, and among species. The quantitative and comprehensive nature of the indices permits policy and landscape planners to compare alternative futures as they work to attain policy objectives.

# Chapter 12

# Impacts on Mammal Communities: Landscape Metrics

## Robert C. Corry

When agricultural landscapes are changed by policy, especially when ecological benefits are the goal, success is often measured by area, such as acreage in the Conservation Reserve Program (CRP), lengths and widths of stream or wetland buffers, or amount of improved habitat (Mitsch et al. 2001; Chapter 3, this volume). In this chapter, I report on results obtained from using measurements of landscape patterns to judge the ecological implications of three alternative policy scenarios for two Iowa study watersheds (see Chapter 4). I used landscape pattern metrics (LPMs) to compare the alternative landscape futures for each scenario in these watersheds and estimate the scenarios' implications for small mammal habitat. I address small mammal habitat here because small mammals are common to Corn Belt landscapes, and they can sustain viable populations amid intensive agriculture—if at least some noncrop habitats remain in the agricultural landscape. By comparing the habitat consequences of different landscape futures, I suggest how small mammal habitat in the Corn Belt may be affected by alternative agricultural policies.

## Small Mammal Habitat Characteristics

Significant biodiversity can exist in small patches within agricultural landscapes (Schwartz and van Mantgem 1997; Ludwig 1999). In the Corn Belt, small patches contribute substantial biodiversity and habitat to the highly fragmented landscape. Small habitats can be encouraged and managed to promote myriad ecological functions, including sequestering carbon, filtering runoff, providing for pollination, and managing pests (Corry and Nassauer 2002). For example, farmers who plant corn with the Bt (Bacillus thuringiensis) gene are required to plant 20 percent of the corn area as non-Bt corn to protect against Bt resistance in crop pests, a pattern:process relationship that incorporates relatively small patches (Pioneer Hi-Bred International Incorporated 2005).

In the highly fragmented agricultural landscapes of the Corn Belt, ecological functions include habitat for animals that tolerate small habitat patches, nearby roads, and agricultural disturbance (e.g., tillage, crop protection, and harvest). To measure some basic habitat potentials of the future scenarios, I chose small mammal habitat because these mammals respond to land cover at the scale of a few meters (1 meter = 1.09 yards). In addition, small mammals have dispersal distances and home-range sizes that (1) fit within the area of the study watersheds; (2) require good habitat dispersed throughout the landscape, as opposed to being concentrated in a few, isolated reserve areas; and (3) comply with the common sizes and shapes of habitat patches in highly fragmented contexts (Foster and Gaines 1991; Peles et al. 1999). For small mammals, the best quality habitats include small and linear patches such as roadsides and fencerows. To allow comparison of results among different biodiversity assessments of the scenarios, I chose small mammal species that are a subset of those given in Chapter 11 (Tables 11-2 and 11-3). This subset included common nonpest species that ranged in size from the plains pocket mouse (*Perognathus flavescens*) to the woodchuck (*Marmota monax;* see Table 12-1).

## Corn Belt Landscape Patterns

The two study watersheds (Chapter 4) are typical of Corn Belt land division and management in the context of intensive agriculture (e.g., square-mile sections and large fields of corn and soybeans). Patches of relatively high biodiversity are small, infrequent, often linear, and at the interstices between large, relatively biologically impoverished patches of crops (Corry and Nassauer 2002). These small patches may provide two important landscape functions: the capture and concentration of scarce resources, and the conservation of a high diversity of organisms (Ludwig 1999). Schwartz and van Mantgem (1997) characterize the common fragmented pattern of the Midwest as relegating substantial biodiversity to small, less than 25 acre (10 ha), isolated patches amid mile-wide crop fields. In these Corn Belt alternative futures, noncrop areas vary in number, size, and shape, but are commonly few, small, and linear. Good-quality

**Table 12-1.** Target Mammal Species and Biological Parameters

| Scientific name | Common name | Home range, acres (hectares) | Dispersal distance, feet (meters) |
|---|---|---|---|
| *At-ground nesting species* | | | |
| **Blarina brevicauda** | Short-tailed shrew | 1.457 (0.590 ) | 196.850 (60) |
| **Cryptotis parva** | Least shrew | 2.964 (1.200) | 885.827 (270) |
| **Microtus ochrogaster** | Prairie vole | 0.124 (0.050) | 1049.869 (320) |
| **Microtus pennsylvanicus** | Meadow vole | 0.099 (0.040) | 918.635 (280) |
| **Peromyscus leucopus** | White-footed mouse | 0.200 (0.081) | 1410.761 (430) |
| **Peromyscus maniculatus** | Deer mouse | 1.482 (0.600) | 1640.420 (500) |
| **Sorex cinereus** | Masked shrew | 0.099 (0.040) | 918.635 (280) |
| **Synaptomys cooperi** | Southern bog lemming | 0.148 (0.060) | 360.892 (110) |
| *Belowground nesting species* | | | |
| **Perognathus flavescens** | Plains pocket mouse | 0.012 (0.005) | 229.659 (70) |
| **Marmota monax** | Woodchuck | 7.657 (3.100) | 984.252 (300) |
| **Zapus hudsonius** | Meadow jumping mouse | 1.482 (0.600) | 1640.420 (500) |

*Source*: Adapted from Clark et al., Chapter 11, this volume.

habitat typically occurs as field boundaries, roadsides, or stream buffers; relatively poor habitat occurs as large expanses of corn, soybeans, alfalfa, and pasture. A few large patches of woodland or perennial cover (e.g., CRP parcels in the baseline Buck Creek watershed) are good habitats. Scenarios 2 and 3 present more heterogeneous landscapes with the intention of more and better connected habitats. The baseline landscape and Scenario 1, on the other hand, were relatively homogeneous and lacked extensive and connected habitats.

## Conducting the Study

I measured habitat quality using LPMs, which quantify differences in the structure of ecosystems (such as "how much" and "where") across mappable areas ranging from yards to miles (meters to kilometers) wide. LPMs are relatively efficient at characterizing landscapes—taking a few minutes to hours for calculation depending on data and number of metrics—which is an attractive part of their utility. Because they are easy and efficient to apply, LPMs may be useful for judging alternative future landscapes for ecological outcomes (Smith et al. 2000).

I used the LPMs to quantify the landscape attributes and judge relative habitat quality. Methods are described in greater detail elsewhere (Corry 2005; Corry and Nassauer 2005). Map (geographic information systems [GIS]) data were reclassified into "good," "moderate," and "poor" habitat classes for two guilds of small mammals that would use habitat differently, those that nest at ground level and those that nest below ground level. LPMs were applied to the GIS data for all landscape scenarios for each of four attributes of landscape pattern that are relevant to habitat function:

- amount of habitat
- landscape heterogeneity (the diversity of habitats available, especially the amount, location, and quality of food and cover)
- average size of good habitat patches
- connectedness of habitat.

## Evaluating the Study Results

Each landscape attribute was quantified with an LPM. LPM values reported here are the best performing metrics based on comparative evaluations (Corry and Nassauer 2005). The units and range of LPMs differ, so for comparison purposes untransformed (raw) LPM values are reported for baseline conditions. For alternative landscape scenarios, LPMs are reported as a percentage of change (direction and magnitude) from baseline conditions (Table 12-2).

### Baseline Landscapes

The proportion of good habitat (as proportion of total landscape area [PLAND]) for the baseline landscapes ranges from a low 4.9 percent value to more than one-quarter of watershed area. For both small mammal guilds, the Buck Creek baseline landscape has relatively large amounts of good habitat (including CRP) at 26.9 percent; the more intensively cropped Walnut Creek landscape has less than 5 percent good habitat proportions. Buck Creek's woodlands are substantial for their contributions to habitat area; Walnut Creek lacks extensive woody cover.

**Table 12-2.** Percentages of Change in LPM Values from Baseline Landscape Conditions for the Three Alternative Landscape Scenarios

| | Buck Creek landscapes | | | | | | Walnut Creek landscapes | | | | | |
| | Scenario 1 | | Scenario 2 | | Scenario 3 | | Scenario 1 | | Scenario 2 | | Scenario 3 | |
| | AG | BG | AG | BG | AG | BG | AG | BG | AG | BG | AG | BG |
|---|---|---|---|---|---|---|---|---|---|---|---|---|
| **Proportion of habitat (PLAND)** | −83 | −83 | −79 | −79 | +118 | −58 | −42 | −42 | −41 | −41 | +762 | +55 |
| **Landscape heterogeneity (MSIEI)** | −6 | −29 | +32 | −26 | +11 | +1 | −58 | −39 | +63 | +240 | +363 | +325 |
| **Habitat patch size (MPS)** | −88 | −88 | −84 | −84 | +114 | −67 | −50 | −50 | −59 | −59 | +979 | +33 |
| **Habitat connectivity (MNN)** | +16 | −5 | 0 | −50 | −3 | −37 | +18 | +35 | +9 | −39 | −36 | −16 |

Notes:
AG refers to at-ground nesting guild of small mammals.
BG refers to below-ground nesting guild of small mammals.
+/− refers to the direction of change from baseline as positive or negative.
Percentage values are rounded to integers.

I measured landscape heterogeneity with the Modified Simpson's Evenness Index (MSIEI), which is a measure of relative distribution of area among all land cover types. In the baseline landscapes, heterogeneity was variable for the small mammal guilds. On a scale from 0 to1, evenness values ranged from a low of 0.134 (very uneven) for at-ground nesting mammals in Walnut Creek, to a high of 0.812 (approaching evenness) for belowground nesting mammals in Buck Creek. For both small mammal guilds, the Walnut Creek landscape was much lower in landscape heterogeneity than the Buck Creek landscape. This makes sense because the distribution of area of land cover types in the baseline Walnut Creek landscape was primarily to row-crop production. Buck Creek land cover types were more evenly proportioned among row crops, pasture and alfalfa, and woodlands.

On average, habitat patch sizes in the Buck Creek baseline landscape were slightly larger than those in Walnut Creek. The mean patch size (MPS) of good habitats in Buck Creek was 3.21 acres (1.3 ha), and Walnut Creek MPS was 2.47 acres (1.0 ha). Considered with the proportion of land cover types, it is clear that the Buck Creek baseline landscape had more habitat in bigger patches, and Walnut Creek had relatively little good habitat area concentrated in smaller patches.

The connectedness of habitats is an important attribute that facilitates the flows of species, materials, and energy through landscapes (Tewksbury et al. 2002). Explicit connectivity (where habitat patches are physically together at boundaries) is impossible in Corn Belt landscapes fragmented by roads, railways, and power lines. For this attribute of landscape pattern, closely clustered patches of habitat were considered to equate to better connectivity (compared to loosely clustered habitat patches). The measure of habitat connectivity is the average distance among good habitat patches, or Euclidean mean nearest neighbor (MNN) distance.

MNN distances for the baseline landscapes ranged from 33.46 feet (10.2 m) (for at-ground nesting mammals in Buck Creek) to 426.18 feet (129.9 m) (for belowground nesting mammals in Walnut Creek). In both watersheds (Buck Creek and Walnut Creek respectively), good habitat for at-ground nesting mammals, on average, was much closer (33.46 and 65.94 feet [10.2 and 20.1 m]) than for belowground nesting mammals (237.53 and 426.18 feet [72.4 and 129.9 m]). For the two small mammal guilds, the average dispersal distance for at-ground nesting species is 921.92 feet (281 m) with a range of 196.85–1,640.42 feet (60–500 m); and 951.44 feet (290 m) with a range of 229.66–1,640.42 feet (70-500 m) for belowground nesting mammals. For many (but not all) of the mammals I considered, these baseline landscapes have good habitat patches within average dispersal distances.

The LPMs showed widely varying amounts of landscape pattern change for the alternative futures, ranging from changes as small as 1 percent to those as large as 979 percent from baseline conditions (Table 12-2). The smallest changes were for connectivity (MNN), indicating that connectivity changed least among the futures compared with other small mammal habitat characteristics. The highest variability among alternative futures was for the size of habitat patches (MPS), particularly in the Walnut Creek watershed for the at-ground nesting small mammals. The implication is that substantial habitat improvement is possible in a landscape that is largely devoted to corn and soybean production under baseline conditions.

## Scenario 1 Landscapes

All Scenario 1 landscapes had reductions in habitat proportions of 42 to 83 percent less than baseline conditions. Habitat patch size also decreased by 50 to 88 percent for Scenario 1 futures compared with the baseline, and landscape heterogeneity decreased by 6 to 58 percent from the baseline. Good habitats were 16 to 35 percent less connected for all but one case (for belowground nesting mammals in Buck Creek, habitats were 5 percent more connected than under baseline conditions). Across almost all measures, the landscapes resulting from Scenario 1 offered poorer habitat conditions for small mammals. In Buck Creek, the Scenario 1 alternative future led to dramatic reductions in belowground nesting mammal habitat area (83 percent less). The few remaining habitat patches were in concentrated areas, and the distance among them was less than among the larger, more evenly distributed habitat patches of the baseline conditions.

## Scenario 2 Landscapes

In the alternative futures of Scenario 2, the proportion of small mammal habitat was lower than that of the baseline, and decreased almost as much as in the alternative futures of Scenario 1. Habitat proportion decreases ranged from 41 to 79 percent less than baseline. The mean habitat patch sizes were also smaller than under the baseline, and for Walnut Creek the patch sizes were smaller than under Scenario 1. Mean patch sizes decreased by 59 to 84 percent from baseline. The MSIEI measure for heterogeneity showed more heterogeneous landscapes by 32 to 240 percent, except in the case of belowground nesting mammal habitats in Buck Creek (which were less heterogeneous by 26 percent). Finally, compared with the baseline, the connectivity of habitats under Scenario 2 was 0 to 9 percent better for at-ground nesting mammals and 39 to 50 percent worse for belowground nesting mammals.

Many of these results run counter to expectations. The most obvious explanation is the conventions used here for habitat classification. Except for the measure of landscape heterogeneity, which accounts for every land cover type and patch in the landscape, these LPMs quantify only *good* habitat patches. Because substantial areas of forage and pasture translated on aggregate to moderate habitat quality, we see a dramatic reduction in good habitat and measures of habitat patch size and connectedness do not account for the possible habitat benefits of moderate habitat. In other words, with so few good habitat patches remaining, they are often much smaller than baseline and much more dispersed in the landscape.

## Scenario 3 Landscapes

The proportions of good habitat in the alternative futures resulting from Scenario 3 increased by 55 to 762 percent for Walnut Creek, and increased for at-ground nesting mammals in Buck Creek by 118 percent. For belowground nesting mammals in Buck Creek, habitat proportions decreased by 58 percent, and mean habitat patch size decreased by 67 percent. Mean habitat patch sizes for all other cases were 33 to 979 percent larger than baseline conditions. Landscape heterogeneity for belowground nesting mammals in Buck Creek increased by less than 1 percent over baseline; others increased by 11 to 363 percent. Good habitat patches for all Scenario 3 landscapes were 3 to 37 percent more connected than under baseline conditions.

The belowground nesting mammals in Buck Creek generally lost habitat quality under Scenario 3, in part because in this future, strip intercropping land cover (with prairie strips) was employed over a large proportion of the rolling Buck Creek watershed. This type of land cover is classified as moderate habitat quality for belowground nesting mammals. Under baseline conditions, the Buck Creek watershed included 16 percent CRP, which is classified as good habitat. As a result, proportions, patch sizes, and connectedness of habitats for belowground nesting mammals were enhanced compared with Scenario 3. Under baseline conditions, 26.9 percent of the watershed was good habitat for belowground nesting mammals, but in Scenario 3 this value drops to 11.3 percent. For all other habitat measures, habitat in Scenario 3 improved by the largest of margins (up to 979 percent improvement over baseline conditions).

## Conclusions from This Study

The LPMs applied here yield some general findings. First, the Walnut Creek and Buck Creek watersheds differ markedly in baseline habitat quality. The Buck Creek baseline landscape measures more than five times as much habitat, in patches that are, on average, 25 and 40 percent more connected than the baseline in Walnut Creek. The Buck Creek baseline landscape is also four times more heterogeneous than that of Walnut Creek. Initial conditions in Buck Creek are relatively good for small mammal diversity and abundance (i.e., more diverse land covers and less proportion of cultivation). As a result, substantial measurable improvements in small mammal habitat in the Buck Creek watershed require much more dramatic landscape changes than in the Walnut Creek watershed. In Walnut Creek watershed, baseline conditions are relatively poor for small mammal diversity and abundance. The watershed is relatively homogeneous and intensively cropped, with more than 83 percent of its area in row crops. Consequently, the future scenarios yield more dramatic habitat improvements. The two watersheds display differences in agricultural land use intensity that are apparent throughout the Corn Belt.

The most consistent results from LPM application were the diminished habitat quality in the alternative future landscapes of Scenario 1 and the improved habitat quality for the landscapes of Scenario 3, compared with the baseline. The alternative futures of Scenario 2 did not, by LPM values, show consistent improvements in habitat quality. On most landscape attributes, the LPM values actually showed decreases in habitat quality for Scenario 2 compared with baseline conditions. For Corn Belt farms, policies that aim to improve biodiversity have beneficial consequences for small mammal habitat.

Some notable limitations of LPM measurements affect these results (described in detail in Corry and Nassauer 2005). For example, where variation in land cover is not represented in GIS data, LPMs do not account for that variation. Because the linear habitats in strip intercropping (especially with prairie strips) were not explicitly represented in the GIS data, the LPM probably underestimated connectivity of good habitats for Scenario 3 alternative futures, in which prairie strips were prominent.

Comparing LPM results to those reported by Clark et al. in Chapter 11 yields some similarities and differences. Their model includes more species and interactions, whereas my LPM application was limited to an aggregation of a few species (eight at-ground nesting small mammals and three belowground nesting small mammals) into guilds without interaction effects. In addition, the dynamic model the researchers used in Chapter 11 explicitly incorporates species competition, predation, and mortality—factors that are not included in the LPMs.

The beginning conditions of Buck Creek's baseline landscape and the relative improvements as measured by LPMs are congruent with the results of the community modeling approach (Chapter 11) in that 80 percent of potential diversity was achieved in all Buck Creek landscapes. In addition, Scenario 3 had the highest absolute diversity. These results suggest that the Scenario 3 policies for improving biodiversity had the desired effect. LPM results do not, however, support Clark et al.'s finding that Walnut Creek's absolute species diversity was greater for the landscapes of Scenario 2 than for the baseline landscapes.

The general trends identified in Chapter 11 are more marked than in my LPM results. For example, my LPM habitat measures for Walnut Creek did not show a similar magnitude of improvement over baseline conditions for the alternative futures of Scenarios 1 and 2, as reported by Clark et al. This may be because LPM values focus primarily on good habitats and moderate habitat classes are not integrated into the reported values. The dynamic modeling approach used by Clark et al. may be more sensitive to habitat contributions from moderately improved habitats, which may be particularly characteristic of small mammal habitats in Corn Belt agricultural landscapes. These moderate quality habitats are especially extensive in landscapes resulting from Scenarios 2 and 3.

## Implications for Agricultural Policy

Generally, the LPMs for four landscape attributes—habitat proportion, habitat patch size, landscape heterogeneity, and habitat connectivity—show that small mammal habitat is of the best quality under the landscapes resulting from Scenario 3. Habitat quality resulting from baseline conditions and Scenario 2 is the next best, with Scenario 1 offering the lowest overall habitat quality. These LPMs imply that agricultural policies can have measurable desired outcomes. Since Corn Belt agriculture is among the most intensive land uses within the Mississippi River Basin (MRB), the improvements quantified by the four landscape attributes considered here could have beneficial effects on small mammal habitat quality across the entire basin.

The LPMs I applied indicate that the policies of Scenario 3 can create landscapes that broadly substantially improve all small mammal habitat attributes compared with any other alternative. In contrast, policies driving Scenario 1 diminish habitat quality. Furthermore, my results generally support those of the spatially explicit population model (Chapter 11), and differences in results can be attributed to some limitations of the LPM technique for small patch habitat landscapes.

*Chapter 13*

# Wildlife Habitat

Kathryn Freemark Lindsay, Mary V. Santelmann,
Jean C. Sifneos, Denis White, and David A. Kirk

One of the most sensitive indicators of ecosystem change is the response of native plant and animal species to alterations in land cover and management practices, and species response is of potential use in ecosystem risk assessments (Pratt and Cairns 1992; White et al. 1999). Here we describe one of several modeling approaches used to evaluate the alternative futures for Iowa watersheds described in Chapter 4: the potential impacts of habitat change on wildlife. In addition to the alternative futures, we also evaluated a reconstructed pre-European settlement landscape for a past perspective on changes in habitat for native species in the study watersheds. Our approach, modified from White et al. (1997), was based on the premise that the level of impact on a species increases as its habitat is depleted or degraded. Our work required a habitat map for each scenario (past, baseline, and alternative futures); a list of resident species; and an estimate of the suitability of each habitat for each of those species. Each scenario's impact on wildlife habitat was calculated as the median percent change in habitat for a given set of species relative to the baseline.

The use of scenario-based futures (UNEP 2003; Millennium Assessment 2005) can help inform decisionmakers about the potential impacts of their decisions, and encourage incorporation of ecological principles into land use decisionmaking (Dale and Haeuber 2001). Our goals were to quantify the effects of habitat conversion on species richness in the future scenarios relative to the baseline landscape and the past, and to explore how these effects are manifested spatially across the landscape.

### Conducting the Study

We studied two watersheds, Walnut Creek and Buck Creek in Iowa. The baseline and past landscapes for these watersheds are described in Chapter 4 and elsewhere (Santelmann et al. 2005). For both watersheds, we included all mammal, bird, reptile, amphibian, and butterfly species found in central Iowa (for details see Santelmann et al. 2005). We eliminated some species from the list and did not model others if they were extirpated or unlikely to be designated free-ranging status (e.g. elk [*Cervus Canadensis*], bison [*Bison bison*], mountain lion [*Felis concolor*], and passenger pigeon [*Ectopistes migratorius*]). If we had included extirpated species in our analysis of past landscapes but not in the futures, the resulting indices would not have been comparable, so we chose

**Table 13-1.** Number of Species in Each Set of Species in the 26 Habitat Classes[a]

| Code | Map class | Habitat class | Native | av.suit | Vertebrate species | | | | Introduced | S1 and S2 | Lepidoptera |
| | | | | | Amphibian | Reptile | Mammal | Bird | | | Butterfly |
|---|---|---|---|---|---|---|---|---|---|---|---|
| 1 | 2 | rc.cp | 91 | 0.66 | 0 | 16 | 24 | 51 | 7 | 5 | 10 |
| 2 | 2 | rc.ct | 103 | 0.9 | 0 | 16 | 36 | 51 | 7 | 7 | 10 |
| 3 | 3 | rc.ns | 122 | 1.19 | 0 | 16 | 44 | 62 | 7 | 12 | 16 |
| 4 | 4 | rc.sg | 69 | 0.58 | 0 | 16 | 24 | 29 | 7 | 7 | 2 |
| 5 | 4 | rc.fall | 61 | 0.67 | 0 | 16 | 41 | 4 | 3 | 7 | 1 |
| 6 | 3 | rc.os | 115 | 1.25 | 0 | 16 | 42 | 57 | 7 | 11 | 12 |
| 7 | 2 | rc.o | 92 | 0.79 | 0 | 16 | 23 | 53 | 7 | 7 | 11 |
| 8 | 7 | farmstead | 110 | 1.36 | 0 | 20 | 36 | 54 | 7 | 8 | 40 |
| 9 | 4 | strip.h | 115 | 1.47 | 0 | 25 | 45 | 45 | 6 | 11 | 21 |
| 10 | 5 | strip.w | 113 | 1.6 | 0 | 23 | 37 | 53 | 4 | 10 | 10 |
| 11 | 4 | grass.crp | 117 | 1.53 | 0 | 26 | 44 | 47 | 5 | 16 | 57 |
| 12 | 1 | grass.hay | 112 | 1.11 | 0 | 27 | 40 | 45 | 5 | 16 | 39 |
| 13 | 4 | grass.pd | 115 | 1.55 | 0 | 27 | 45 | 43 | 5 | 18 | 81 |
| 14 | 6 | grass.pw | 128 | 1.91 | 10 | 26 | 43 | 49 | 5 | 16 | 76 |
| 15 | 4 | shrub.past | 127 | 1.4 | 2 | 27 | 44 | 54 | 6 | 16 | 53 |
| 16 | 4 | shrub.ung | 128 | 1.71 | 2 | 27 | 43 | 56 | 5 | 11 | 55 |
| 17 | 5 | for.50 | 125 | 1.68 | 0 | 27 | 33 | 65 | 4 | 8 | 36 |
| 18 | 5 | for.rug | 171 | 2.29 | 7 | 27 | 34 | 103 | 3 | 12 | 40 |
| 19 | 5 | for.upug | 140 | 1.86 | 0 | 26 | 32 | 82 | 3 | 8 | 35 |
| 20 | 5 | for.upg | 125 | 1.56 | 0 | 26 | 32 | 67 | 3 | 8 | 27 |
| 21 | 5 | for.sug | 100 | 1.39 | 0 | 27 | 47 | 26 | 3 | 11 | 64 |

Table 13-1.  Number of Species in Each Set of Species in the 26 Habitat Classes[a] (continued)

| Code | Map class | Habitat class | Native | av.suit | Vertebrate species | | | | | | Lepidoptera |
| | | | | | Amphibian | Reptile | Mammal | Bird | Introduced | S1 and S2 | Butterfly |
|---|---|---|---|---|---|---|---|---|---|---|---|
| 22 | 5 | for.sg | 94 | 1.14 | 0 | 27 | 42 | 25 | 3 | 12 | 64 |
| 23 | 6 | wet.sp | 111 | 1.5 | 12 | 26 | 11 | 62 | 2 | 10 | 26 |
| 24 | 6 | wet.pond | 92 | 1.39 | 12 | 27 | 13 | 40 | 0 | 9 | 14 |
| 25 | 6 | wet.st | 63 | 1.05 | 7 | 28 | 13 | 15 | 0 | 8 | 33 |
| 26 | 6 | wet.eng | 90 | 1.11 | 8 | 26 | 12 | 44 | 0 | 6 | 24 |
| | | Total | 239 | na | 12 | 29 | 52 | 146 | 8 | 24 | 117 |

[a]Row crop chisel plow, row crop conservation tillage, row crop native strip, row crop small grains, row crop fallow, row crop organic strip, row crop organic, farmstead, herbaceous strip, woody strip, Perennial cover non-crop, CRP, alfalfa/hay, dry prairie, wet prairie, pasture, ungrazed shrubland, ungrazed forest less than 50 years old, ungrazed riparian forest, ungrazed upland forest, grazed upland forest, ungrazed savanna, grazed savanna, semipermanent wetland, pond, stream, and engineered wetland.

Notes: The seven map classes (alfalfa, row crops, strip intercropping, perennial herbaceous cover, woodland/woody cover, water/wetland, and urban/residential/roads) into which the habitat classes are mapped in Figures 13-1 through 13-2 and Color Figures 47-49 are shown in Column 2, and the average suitability score for each habitat class (averaged over all native vertebrate species) is shown in Column 5. The abbreviation av.suit stands for average suitability.

to omit extirpated species from calculations of the habitat change statistics and maps of species richness. The limitation of this approach is that it underestimates the effect of historical changes on species richness in these landscapes.

By reviewing bird species composition and abundance among habitats using information from Best et al. (1995) and then evaluating potential differences for other vertebrates, we derived 26 wildlife habitat classes (Table 13-1) from land cover classes in our geographic information system (GIS) coverage. By reviewing the literature and obtaining expert opinions, we ranked habitat suitability for all species in each potential habitat type and generated a draft matrix of species–habitat associations. The literature sources used included Jackson et al. (1996) and Kent and Dinsmore (1996). For each taxonomic group, we then asked two to four additional experts for each taxonomic group to review the draft matrix (see Santelmann et al. [2005] for the complete matrix, the list of reviewers, and further methodological details).

We included 239 native vertebrate species (146 birds, 52 mammals, 29 reptiles, and 12 amphibians); 8 introduced vertebrate species; and 117 butterfly species in our analysis. Of all habitat classes, ungrazed riparian and upland forest had the highest species richness with 171 and 140 species, respectively. Few species were found in only one of the 26 habitat classes; semipermanent wetland and ungrazed riparian forest had the most unique species at 5 and 4, respectively.

For nonbird taxa lacking comprehensive studies, reviewers considered a 0- to 4-point scale to be appropriate. We applied this scale to represent habitat suitability as follows: 0 = unsuitable habitat in which the species is never found, except in transit; 1 = sink or marginal habitat of the lowest quality, insufficient to sustain a population; 2 = sink or marginal habitat of sufficient quality to maintain a population temporarily, and/or with immigration from higher quality habitat; 3 = source habitat capable of supporting individuals during those life stages most critical for sustaining populations for extensive periods without immigration; and 4 = source or optimal habitat with the highest reproduction and/or survival rates capable of sustaining populations indefinitely. The matrices of species–habitat associations are available from the authors (U.S. EPA 2006).

We used the habitat associations derived for sets of species within taxonomic groups to map habitat for each species in the past, at a baseline, and at each future scenario with and without crop rotation. Each GIS map consisted of the score for a species in the habitat at each pixel (3 m [9.8 ft] resolution) location. From these maps of habitat scores we estimated the total amount of potential habitat for a species as the sum of all the scores across the watershed for each alternative future, the baseline landscape, and the past. We then calculated the percentage change in potential habitat for each species relative to the baseline landscape (*present* in Equation 13-1) for the three future scenarios and for the past. Finally, we used the median of the percentage changes for different groups of species as a summary statistic, following White et al. (1997).

The formula for calculating the habitat change score, $HC_j$, for a specific group of species for one of the future landscapes (or for the past landscape) was

$$HC_j = median \left[ \sum_i^{species} (hab_{i,j} - hab_{i,present})/hab_{i,present} *100 \right], \quad (13\text{-}1)$$

for a future or past landscape, $j$, where $hab_{i,j}$ is the suitability-weighted abundance of species $i$ in the future or past landscape $j$, and $hab_{i,present}$ is the suitability-weighted abundance of species $i$ in the baseline present landscape.

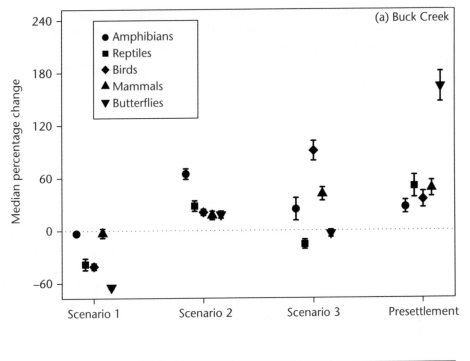

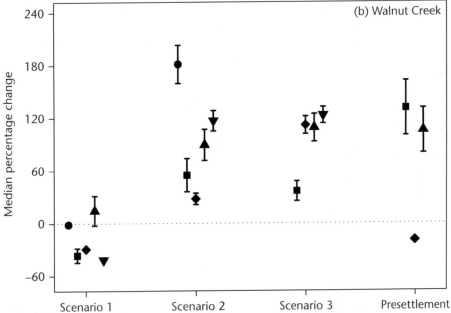

**Figure 13-1.** Median Percent Change in Habitat Area for Taxa of Native Species for the Two Watersheds

*Notes:* Standard deviation, ±1. Habitat area adjusted by suitability compared to the baseline. Taxa exclude introduced, extinct, and extirpated species. Values >0 indicate habitat gain compared to the baseline; values <0 indicate habitat loss compared to the baseline. Species groups with changes greater than 200 percent are not shown (Walnut Creek amphibians in Scenario 3 and Presettlement and butterflies in Presettlement). See Table 13-3 for values.

When the median percent change statistic was positive it meant that more habitat for the species occurred in the watershed in the future or past landscape than in the baseline landscape; negative values meant that less habitat occurred in the watershed in the future or past landscape than in the baseline landscape. We grouped sets of species as follows: native birds, mammals, reptiles, amphibians, and butterflies; all native vertebrates; all introduced vertebrates; and all rare vertebrates. We designated rare vertebrates as those with state conservation ranks of S1 or S2 (rare and vulnerable, S1 is most severely threatened and S2 is highly vulnerable but less so than S1), as determined by the Iowa Natural Areas Inventory (Iowa DNR 2006).

To investigate the effects on the habitat change statistics of possible errors in the species–habitat suitability scores, we performed a Monte Carlo simulation study. We altered the suitability scores under an assumed error model (normally distributed errors with a mean of zero and a standard deviation (sd) of an integral number of score levels) and variability in the results computed. This error study primarily addressed the suitability of a habitat for a species, not whether the species was present or absent. We generated a species–habitat suitability matrix containing a modified suitability score for each species in each habitat. We then modified scores by combining a term from the error model with the original scores. If the resulting score was less than zero we set it to zero, if greater than six for birds we set it to six, or if greater than four for other species we set it to four. Scores originally set to zero were maintained at zero and not altered. Consequently, a species could change from present to absent if the score became zero, but could never change from absent to present. Errors were generated in this way and the habitat statistics were calculated for each taxonomic group as well as introduced and rare vertebrate species subsets. We repeated the error generation process 1,000 times, and the mean and standard deviation of the median statistics were calculated.

We mapped vertebrate species richness for the alternative futures and the past compared to the baseline landscape. For each species we created a presence/absence map from the suitability-weighted habitat map by modifying all pixel scores greater than or equal to one to "present" and all scores of zero to "absent." For each future and the past, the number of species present in each pixel of habitat for the alternative future or past was subtracted from the number of species present in the same pixel in the baseline landscape. A positive number indicated a gain in species richness, while a negative number indicated a loss in species richness for that pixel of habitat in the future or past, compared to the baseline landscape.

## Evaluating Our Results

Changes in habitat area relative to the baseline landscape for butterflies and for vertebrates by taxon (Figure 13-1) and for vertebrates by species of concern (Figure 13-2) varied between watersheds and among scenarios (Table 13-2). Variability in mean percent change in habitat area (as estimated from modeled uncertainty in habitat suitability scores) was generally less than 20 percent (±1 sd) except for amphibians and introduced species, which were more variable because of the smaller number of species in these taxonomic groups (Table 13-3).

More habitat was available in the past relative to the baseline landscape in Buck Creek for all native taxa (27 ± 8 to 164 ± 17 percent; Figure 13-1a). Similarly, in Walnut Creek (Figures 13-1b and 13-2b), native vertebrates overall and most taxa (particularly amphibians and butterflies) had more habitat in the past than in the baseline

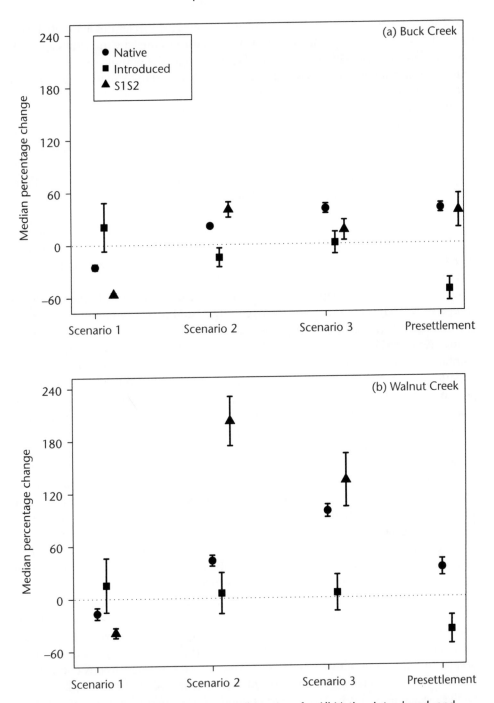

**Figure 13-2.**    Median Percent Change in Habitat Area for All Native, Introduced, and Rare (S1 and S2) for the Two Watersheds

*Notes:* Standard deviation, ±1. Habitat area adjusted by suitability compared to the baseline. Values >0 indicate habitat gain compared to the baseline; values <0 indicate habitat loss compared to the baseline. Results for S1 and S2 species for Presettlement in Walnut Creek are not shown because values were much greater than 200 percent. See Table 13-3 for values.

landscape (34 ± 10 to 11,529 ± 1,425 percent). Native bird species not associated with wetlands, however, had less habitat in the past relative to the baseline landscape (–19 ± 0.1 percent) because semipermanent and ephemeral prairie pothole wetlands (which were previously abundant), have been drained or converted to permanent farm ponds in the baseline landscape in Walnut Creek, creating more upland habitat in the baseline landscape. Also, we note that the percent change statistic does not include species associated only with wetland habitats because such species have no habitat in the baseline landscape (and therefore the change statistic is undefined). In the past, however, wetland-associated species would have had extensive habitat in the Walnut Creek watershed. A study by Rustigian et al. (2003) indicates that the presence of prairie pothole wetlands greatly increased modeled population sizes for amphibian species in Walnut Creek watershed.

Not surprisingly, less habitat was available for introduced species in the past relative to the baseline in both watersheds. In contrast, rare species (S1 and S2) had more habitat in the past relative to the baseline in both watersheds, but particularly in Walnut Creek (Figure 13-2).

Under Scenario 1 (Chapter 4), habitat for all taxa declined in both watersheds compared with the baseline landscape (Figure 13-1) and native vertebrates habitat declined overall (Figure 13-2; –1 ± 0.2 to –65 ± 2 percent, respectively). Although habitat change for introduced species was similar among all future scenarios, rare species lost habitat in both watersheds in Scenario 1 (Figure 13-2). Mammals were the exception, remaining about the same in Buck Creek (–3 ± 5 percent) and Walnut Creek (15 ± 17 percent). This was largely because Scenario 1, like the other future scenarios, specified a no-till management practice in corn–soybean row crops, which provided more cover for small mammals than baseline conventional-till practices.

Under Scenario 2, more habitat was available for all taxa (Figure 13-1) and native vertebrates overall (Figure 13-2) in both Buck Creek (17 ± 5 to 65 ± 6 percent, respectively) and Walnut Creek (28 ± 7 to 181 ± 22 percent, respectively) relative to the baseline. Habitat for introduced species declined in Buck Creek, but remained about the same in Walnut Creek. Rare species gained habitat in both watersheds (Figure 13-2).

Under Scenario 3, compared with the baseline, more habitat was available in Walnut Creek for all taxa (Figure 13-1) and native vertebrates overall (Figure 13-2; 37 ± 11 to 1,617 ± 128 percent, respectively). More habitat was available in this scenario for most taxa in Buck Creek (ranging from 24 ± 13 to 91 ± 11 percent), with the exception of reptiles (–16 ± 5 percent) and butterflies (–4 ± 4 percent). Compared with baseline conditions, the decline in habitat area for the latter taxa in Buck Creek can be attributed to conversion of abundant pasture, alfalfa, and Conservation Reserve Program (CRP) land in the baseline year (1994). Under Scenario 3, strip intercropping accounts for the decline in reptile and butterfly habitat. Patches of perennial herbaceous cover in the baseline year (e.g., alfalfa, pasture, and CRP) are assigned as habitat for 27, 27, and 26 reptile species, respectively. The strip intercropping in Scenario 3 is assigned as habitat for only 16 reptile species (Table 13-1). Likewise, more butterfly species are associated with alfalfa, pasture, and CRP (39, 53, and 57 species, respectively) than with strip intercropping (16 species). Overall, habitat for introduced species remained about the same. Rare species gained habitat in both watersheds under Scenario 3 (Figure 13-2).

For most taxa, estimated changes in the statistic for estimating wildlife habitat in Scenarios 2 and 3 (compared to the baseline) were similar to each other and to those for the past, indicating greater habitat than in the baseline landscape or under Scenario 1. For most species (native vertebrates overall, birds, amphibians, and rare species in

**Table 13-2.** Transition Matrix Showing Habitat Conversions[a] and Effect on Species Richness of Native Vertebrates in the Two Watersheds

| Scenario | Habitat type in the baseline | Habitat type in the future or past | Change in area (%) Buck Creek | Change in area (%) Walnut Creek | Change in number of species |
|---|---|---|---|---|---|
| 1 | Upland ungrazed | Row crop conservation till | 3.78 | 1.16 | −37 |
| 1 | Pasture | Row crop conservation till | 10.77 | 2.21 | −24 |
| 1 | Pasture | Alfalfa/hay | 2.59 | na | −15 |
| 1 | CRP/Perennial non-crop | Row crop conservation till | 12.68 | na | −14 |
| 1 | CRP/p.n-c | Alfalfa/hay | 2.86 | na | −5 |
| 1 | Herbaceous | Row crop conservation till | 1.98 | 1.57 | −12 |
| 1 | Alfalfa/hay | Row crop conservation till | 7.96 | 2.52 | −9 |
| 1 | Farmstead and urban | Row crop conservation till | na | 1.27 | −7 |
| 1 | Row crop chisel plow | Row crop conservation till | 42.86 | 80.75 | 12 |
| 2 | Upland ungrazed | Upland grazed | 5.56 | 2.35 | −15 |
| 2 | Row crop chisel plow | Row crop conservation till | 10.79 | 53.61 | 12 |
| 2 | Alfalfa/hay | Pasture | 3.49 | 1.14 | 15 |
| 2 | Row crop chisel plow | Alfalfa/hay | 22.89 | 7.26 | 21 |
| 2 | Row crop chisel plow | CRP | 3.27 | 5.71 | 26 |
| 2 | Row crop chisel plow | Pasture | 3.26 | 12.88 | 36 |
| 2 | Row crop chisel plow | Herbaceous | 1.16 | na | 24 |
| 2 | Pasture | Alfalfa/hay | 1.67 | na | −15 |
| 2 | Savanna un-grazed | Savanna grazed | 1.05 | na | −6 |
| 2 | CRP/peren.n-c | Alfalfa/hay | 11.54 | na | −5 |
| 2 | CRP/peren.n-c | Pasture | 2.08 | na | 10 |
| 2 | Herbaceous | Alfalfa/hay | 1.18 | na | −3 |
| 2 | Pasture | Riparian ungrazed | 1.21 | na | 44 |
| 2 | Row crop chisel plow | Riparian ungrazed | 1.55 | na | 80 |
| 3 | Row crop chisel plow | Row crop conservation till | na | 32.14 | 12 |
| 3 | Row crop chisel plow | Organic strip | na | 2.73 | 24 |

**Table 13-2.** Transition Matrix Showing Habitat Conversions[a] and Effect on Species Richness of Native Vertebrates in the Two Watersheds (continued)

| Scenario | Habitat type in the baseline | Habitat type in the future or past | Change in area (%) Buck Creek | Change in area (%) Walnut Creek | Change in number of species |
|---|---|---|---|---|---|
| 3 | Row crop chisel plow | Prairie dry | na | 1.47 | 24 |
| 3 | Row crop chisel plow | Row crop native strip | 34.95 | 35.07 | 31 |
| 3 | Row crop chisel plow | Upland grazed | 1.25 | 1.8 | 34 |
| 3 | Row crop chisel plow | Prairie wet | na | 3.02 | 37 |
| 3 | Row crop chisel plow | Riparian ungrazed | 4.88 | 3.36 | 80 |
| 3 | Pasture | Prairie dry | 1.41 | na | −12 |
| 3 | Pasture | Row crop native strip | 7.74 | na | −5 |
| 3 | Pasture | Upland grazed | 1.08 | na | −2 |
| 3 | Pasture | Riparian ungrazed | 2.64 | na | 44 |
| 3 | CRP/p.n-c | Row crop native strip | 10.91 | na | 5 |
| 3 | CRP/p.n-c | Upland grazed | 1.04 | na | 8 |
| 3 | CRP/p.n-c | Riparian ungrazed | 2.77 | na | 54 |
| 3 | Herbaceous | Row crop native strip | 1.41 | na | 7 |
| 3 | Alfalfa/hay | Row crop native strip | 6.48 | na | 10 |
| PS | Upland ungrazed | Prairie dry | 1.74 | 1.62 | −25 |
| PS | Pasture | Prairie dry | 5.32 | 1.68 | −12 |
| PS | CRP/p.n-c | Prairie dry | 5.29 | na | −2 |
| PS | Alfalfa/hay | Prairie dry | 3.61 | 1.51 | 3 |
| PS | Farmstead and urban | Prairie dry | na | 2.17 | 5 |
| PS | Herbaceous | Prairie wet | na | 1.46 | 13 |
| PS | Pasture | Upland ungrazed | 8.31 | na | 13 |
| PS | CRP/p.n-c | Upland ungrazed | 10.63 | na | 23 |
| PS | Alfalfa/hay | Upland ungrazed | 5.32 | na | 28 |
| PS | Herbaceous | Upland ungrazed | 1.39 | na | 25 |
| PS | Row crop chisel plow | Upland ungrazed | 12.23 | na | 49 |
| PS | Row crop chisel plow | Semipermanent wetland | na | 3.34 | 20 |

**Table 13-2.**  Transition Matrix Showing Habitat Conversions[a] and Effect on Species Richness of Native Vertebrates in the Two Watersheds (continued)

| Scenario | Habitat type in the baseline | Habitat type in the future or past | Change in area (%) Buck Creek | Change in area (%) Walnut Creek | Change in number of species |
|---|---|---|---|---|---|
| PS | Row crop chisel plow | Prairie dry | 30.35 | 39.28 | 24 |
| PS | Row crop chisel plow | Prairie wet | na | 38.56 | 37 |

[a] Expressed as change in percentage of watershed area from the baseline to each alternative future scenario or the presettlement past (PS = Past)

*Notes*: Negative differences mean that the habitat type in the baseline results has higher species richness than the habitat in the future or past."p. n-c" = Perennial non-crop. "na" means not applicable, the land cover didn't exist in this landscape.

both watersheds; and for mammals in Buck Creek and butterflies in Walnut Creek), Scenario 3 ranked highest in providing habitat, followed by Scenario 2, then Scenario 1. For mammals in Walnut Creek, Scenarios 2 and 3 were equivalent, and both ranked higher than Scenario 1. For reptiles and butterflies in both watersheds, Scenario 2 was similar to or slightly better than Scenario 3, and both ranked higher than Scenario 1. The future scenarios were generally similar to the baseline with respect to habitat area for introduced species.

Species richness mapping for the baseline (Color Figure 47) revealed "hot spots" for native vertebrates in the riparian forests and perennial herbaceous cover (which occur in very different configurations in each of the watersheds) and "cold spots" (i.e., areas of low species richness) in row crops. Suitability scores also tended to be lower for row crop habitats.

Effects of habitat change on native vertebrates for the past and futures relative to the baseline were evident in the species richness difference maps for Buck Creek watershed (Color Figure 48) and Walnut Creek watershed (Color Figure 49).

Past habitat in each watershed supported higher species richness over more area than in the baseline. This difference in species richness was largely because land in row crops in baseline landscapes had been wet and dry prairie in Walnut Creek and dry prairie and upland forest in Buck Creek (Table 13-2) in the past. Species richness was lower in the past compared to the baseline (i.e., dark red) in areas that had been dry prairie in the past but were converted to pasture and ungrazed upland woodland in the baseline landscape (Color Figures 48 and 49).

In Scenario 1, species richness decreased over much of Buck Creek relative to the baseline (Color Figure 48), primarily because woodland, alfalfa/hay, pasture, and CRP set-asides were converted to row crops (Table 13-1). Gains in species richness in both watersheds were mostly derived from converting conventional-till row crop to conservation-till row crop, a change that was particularly extensive in Walnut Creek (Color Figure 49). In both watersheds, Scenarios 2 and 3 supported greater increases in species richness and lower species richness losses over more area relative to the baseline than Scenario 1. In Buck Creek (Color Figure 48), Scenario 3 resulted in increased area with gains in species richness. The conversion of conventional-till row crop, alfalfa/hay, and CRP to no-till and native-strip row crop increased species richness, as did conversion of row crop in riparian areas to wet prairie and forest. Species richness declined in some areas because pasture was converted to native strip row crop. The bioreserve in Scenario 3 for Buck Creek (Color Figure 48) showed a loss of species richness in some areas, because pasture, alfalfa/hay, and CRP lands in the baseline were coded as suitable for more bird species (54, 45, and 47 species, respectively; see Table 13-1) than savanna and dry native prairie (coded as habitat for 26 and 43 bird species, respectively). Savanna and dry prairie were the land cover types restored in the Buck Creek bioreserve in Scenario 3.

The greater number of animal species associated with pasture (but not prairie) is in part an artifact of the classification scheme, which separated restored prairie into wet prairie and dry prairie classes in the species–habitat association matrix, but included all species associated with pasture in a single class. All the species associated with pasture across varying moisture conditions are combined in the pasture land cover class because research on pasture habitats is less complete. In addition, it was possible for reviewers to distinguish species associated specifically with wet or dry prairie based on the larger, richer literature and on surveys for prairie habitats (e.g., Best et al. 1995).

In Walnut Creek (Color Figure 49), the area of species richness gains was about the same in Scenarios 3 and 2, primarily resulting from conversion of conventional-till to no-till row crop in Scenario 2 and to no-till and row crop native strip in Scenario 3 (Tables 13-1 and 13-2) although the resulting landscape configuration was quite different between scenarios.

## Conclusions from Our Study

Our results show that ongoing agricultural intensification, as envisioned under Scenario 1, will lead to further decline of wildlife from habitat loss in farmland. Cropping and management practices envisioned in Scenarios 2 or 3, however, would both potentially benefit wildlife. Analysis of bird count data from six Iowa watersheds (including Buck Creek and Walnut Creek) has demonstrated that (1) bird species richness and abundance are greater in heterogeneous watersheds (such as Buck Creek) than in more homogeneous watersheds dominated by row crops (such as Walnut Creek); (2) significant differences were observed between watersheds according to pairwise comparisons using Monte Carlo permutation tests in canonical correspondence analysis (CCA); and (3) greater variation was seen between plots within heterogeneous watersheds than within homogeneous watersheds (Freemark et al. unpublished data).

Four major points emerged from the various assessment approaches used in Chapters 9–12 of this volume. First, the landscape pattern and habitat present in Scenario 3 was most beneficial to nearly all taxa. For mammals, although approximately 80 percent of the potential diversity was achieved in all scenarios in Buck Creek, absolute diversity was highest in Scenario 3 (see Chapter 11). Absolute mammal species diversity was higher in Scenarios 2 and 3 than in Scenario 1 or the baseline for Walnut Creek. Using landscape pattern metrics, Chapter 12 reports similar results, ranking Scenario 3 as best for mammal diversity. In the case of amphibians, Chapter 10 reports that Scenario 3 had consistently enhanced breeding numbers because of increased availability of high-quality breeding habitat, fewer roadside ditches (which tend to act as sinks), and less isolated breeding sites. Similarly, land use changes in Scenario 1 had a negative effect

**Table 13-3.** The Means and Standard Deviations of 1,000 Replicates of Monte Carlo Estimates of Median Percent Change in Suitability-Weighted Habitat for Selected Groups of Species

| Group | Total Nspp | Scenario 1 | Std. dev. | Scenario 2 | Std. dev. | Scenario 3 | Std. dev. | Presettlement | Std. dev. |
|---|---|---|---|---|---|---|---|---|---|
| *Buck Creek* | | | | | | | | | |
| **Amphibians** | 12 | -3.2 | 0.3 | 64.6 | 6.0 | 24.2 | 12.9 | 27.0 | 7.6 |
| **Reptiles** | 29 | -38.3 | 6.6 | 27.9 | 6.0 | -15.7 | 5.4 | 50.3 | 12.7 |
| **Birds** | 146 | -40.9 | 3.6 | 20.7 | 3.5 | 90.7 | 11.2 | 35.3 | 9.3 |
| **Mammals** | 52 | -3.0 | 5.2 | 17.1 | 4.8 | 41.5 | 7.5 | 47.8 | 9.2 |
| **Native Vertebrates** | 239 | -25.1 | 3.4 | 21.4 | 2.3 | 40.6 | 5.5 | 40.6 | 5.4 |
| **Exotic Vertebrates** | 8 | 20.8 | 27.6 | -14.5 | 10.7 | 1.3 | 12.4 | -52.6 | 13.0 |
| **S1 and S2** | 24 | -56.4 | 1.6 | 39.9 | 8.8 | 15.8 | 11.7 | 37.1 | 19.3 |
| **Lepidoptera** | 117 | -64.6 | 1.9 | 18.0 | 4.0 | -3.5 | 4.1 | 164.0 | 17.4 |
| *Walnut Creek* | | | | | | | | | |
| **Amphibians** | 12 | -1.2 | 0.2 | 181.3 | 21.7 | 1617.3 | 128.1 | 11529.3 | 1425.4 |
| **Reptiles** | 29 | -36.6 | 8.1 | 54.9 | 19.0 | 36.5 | 11.4 | 131.1 | 31.3 |
| **Birds** | 146 | -29.5 | 1.2 | 27.8 | 6.5 | 111.7 | 10.1 | -19.4 | 0.1 |
| **Mammals** | 52 | 14.5 | 16.8 | 89.3 | 17.7 | 108.7 | 15.9 | 105.5 | 25.7 |
| **Native Vertebrates** | 239 | -16.8 | 6.7 | 42.9 | 6.2 | 98.9 | 7.2 | 33.8 | 9.6 |
| **Exotic Vertebrates** | 8 | 15.5 | 30.9 | 5.9 | 23.4 | 5.5 | 20.7 | -37.2 | 16.2 |
| **S1 and S2** | 24 | -39.1 | 5.7 | 202.1 | 28.1 | 133.8 | 30.3 | 723.9 | 130.6 |
| **Lepidoptera** | 117 | -41.6 | 0.9 | 116.8 | 11.7 | 123.2 | 9.4 | 1189.9 | 105.6 |

*Notes:* The first column shows the total number of species in each group ("Total Nspp"), followed by four sets of two columns that correspond to the three future landscapes and the past landscape. The first and second columns in each set are the mean and standard deviation, respectively, of the median percent change for the group.

on butterfly populations, whereas those in Scenarios 2 and 3 tended to both benefit rare species and increase species richness (see Chapter 9).

Second, different taxa responded differently to components of the different landscape futures. This underscores the need to consider species-specific life history requirements and examine implications of land use change for some sets of species. For example, in Buck Creek, reptiles and butterfly species actually have more habitat in Scenario 2 than in Scenario 3 because wetlands are created at the outlets of tile drains and at road crossings in Scenario 2.

Third, landscape configuration and structure, including the availability and adjacency of different habitat features, can have a profound effect on biodiversity. For example, even in Scenario 3 with biodiversity as its leading goal, in Buck Creek prairie butterfly species benefited under Scenario 3, prairie/wetland species benefited under both Scenarios 2 and 3, and wetland species benefited under Scenario 2. This was because the Buck Creek baseline had more pasture, grassland, and upland than Walnut Creek, where the Scenario 3 design included a prairie/wetland reserve, and overall land cover and practices benefited all butterfly groups (see Chapter 9).

Finally, it is important to point out that some modeled results were counterintuitive. For example, we did not anticipate the finding that—for mammals—the inclusion of interspecific interactions indicated that competition and predation were more intense in larger mammal communities associated with bigger patches of habitat (see Chapter 11). This competition and predation could result in decreased relative density and smaller populations of some species in large patches. Future landscapes will need to be carefully designed, monitored, and managed to ensure that the intended outcomes are realized.

## Implications for Agricultural Policy

This study demonstrates how alternative futures can be used to help explore and communicate the environmental changes and impacts that could result from different policy scenarios. Agricultural intensification, as envisioned under Scenario 1, is likely to lead to further decline of wildlife from habitat loss in farmland. But cropping and management practices such as those envisioned in Scenarios 2 and 3 would potentially benefit wildlife. Public investment in maintaining ecological services in agroecosystems as envisioned in these scenarios could help to conserve native species in agricultural landscapes and to maintain ecologically sustainable agriculture in the future.

## Acknowledgments

This research was supported by grant #R825335-01-0 from the Water and Watersheds Program of the U.S. Environmental Protection Agency (U.S. EPA). Kathryn Freemark Lindsay was supported under contract to Oregon State University. This paper has been subjected to the U.S. EPA's peer and administrative review and approved for publication. Approval does not signify that the contents reflect the views of the agency, nor does mention of trade names or commercial products constitute endorsement or recommendation for use.

# PART 3

*Policy Implications
Across Scales:
From Iowa Watersheds
to the
Mississippi River Basin*

## Chapter 14

# An Integrated Assessment of Alternative Futures for Corn Belt Agriculture

M. Santelmann, D. White, K. Lindsay, J. Nassauer,
J. Eilers, K. Vaché, B. Danielson, R. Corry, M. Clark,
S. Polasky, R. Cruse, J. Sifneos, H. Rustigian,
C. Coiner, J. Wu, and D. Debinski

The impact of agriculture on the earth's ecosystems is unparalleled by any other land use in its spatial extent and intensity of influence (Matson et al. 1997). Humans have had devastating impacts on global biodiversity and biogeochemical cycles (Puckett 1994; Alexander et al. 1996; Vitousek et al. 1997a, 1997b; Sala et al. 2000), and changing agricultural patterns and practices could help to ameliorate those impacts (OECD 2002). We have envisioned alternative futures for agriculture in one of the most productive and intensively cultivated regions of the world, the U.S. Corn Belt. Our results suggest that federal policy could promote ecological, social, and economic well-being at the same time as it reduces the environmental impacts of agriculture on terrestrial and aquatic ecosystems and cities downstream, from the upper reaches of the Mississippi River to the Gulf of Mexico.

This integrated assessment measures outcomes addressing multiple goals in the three alternative scenarios for Corn Belt agricultural landscapes in 2025 (Santelmann et al. 2004; Part 2 of this book). The scenarios maintain agricultural production while changing agricultural land use and management to improve water quality and restore native biodiversity. They aim to introduce change that is acceptable to both farmers and the broader public. Finally, they suggest that the future can be different from the past. Scenario 1 raises the question: "What could these landscapes be like in 25 years with continued priority given to corn and soybean production?" But Scenarios 2 and 3 go further, asking "What *should* they be like?", if policy goals include clean water, diverse native vegetation, abundant wildlife, and attractive rural landscapes populated by farmers and others who appreciate a high quality of life in rural America.

The alternative policy scenarios give varying levels of priority to agricultural production, water quality, and maintenance and restoration of native biodiversity. Chapter 4 describes in detail the alternative future landscapes (that is, "alternative futures") that could result from implementing three alternative policy scenarios in two study watersheds in Iowa (Buck Creek and Walnut Creek). Here, we focus on the interdisciplinary evaluation and integrative assessment of the alternative futures for those study watersheds. We compare the potential for the alternative futures to achieve multiple goals by summarizing and integrating their quantitative impacts for multiple endpoints, which range from economic and social metrics to modeled estimates of water quality and native biodiversity.

These alternative futures are not intended to be prescriptions for specific farms or policy initiatives. Instead, they are intended to inform and inspire decisionmakers to look beyond the existing landscape and envision greater possibilities for public benefits from agriculture (Nassauer and Corry 2004). The alternative futures were designed to allow the exploration of multiple impacts of recommended strategies for addressing social, economic, and environmental effects associated with agriculture, using spatially specific models and computer-simulated images of the resulting landscapes.

Although scientific research can and often does focus on a single facet of a problem, ultimately, landscapes respond in multifaceted ways to landscape change, whether the drivers of change are single goals (such as decreasing the extent of the dead zone in the Gulf of Mexico through reductions in nonpoint source water pollution) or multiobjective goals such as those in the Corn Belt scenarios in this book. A forested riparian buffer may be planted to improve water quality; however, it can also provide habitat for wildlife, reduce erosion, store carbon, and increase the attractiveness of the landscape. A large prairie/wetland reserve may be restored to provide critical habitat for endangered species, and it may also be important for helping recharge regional aquifers with uncontaminated water and in serving as a refuge for a diverse set of native pollinator species. Agricultural practices may be adopted to reduce soil erosion, and also help to meet carbon storage and soil and water quality goals. Because changes in land use affect many processes within an ecosystem, their potential impacts and benefits should be measured across multiple endpoints, requiring insight from scientists in multiple disciplines. This integrated assessment of the alternative futures exemplifies this approach, combining the results from multiple disciplinary teams, each assessing the same landscape for a different type of benefit, each contributing to measuring the overall effect of changes in land use and management on the landscape.

## Study Watersheds and Future Policy Scenarios

The two watersheds chosen for this study, Walnut Creek (13,800 acres [5,130 ha]) and Buck Creek (21,700 acres [8,820 ha]; Color Figures 30–37), were selected to demonstrate the way each scenario might be implemented in different physiographic regions with different sets of agricultural enterprises. Watersheds of the second order were selected because research and decisionmaking at both smaller and larger scales could be informed by investigating responses to land management at this scale. Current land cover for the study watersheds (hereafter termed the baseline landscape) was digitized into a geographic information system (GIS) database from 1:20,000 aerial photographs taken in 1990, and ground-truthed in 1993 and 1994 (Bergin et al. 2000). The study watersheds and three alternative future scenarios (Color Figures 38–43) are described in detail in Chapter 4. The scenarios exemplify what could happen if normative principles were used to achieve societal goals of water quality improvements, biodiversity restoration, and healthy rural communities. If the practices embodied in Scenarios 2 and 3 were used throughout the Upper and Middle Mississippi River basins, particularly in combination with the basin-scale wetland restoration and river management efforts described by Mitsch et al. (2001), we could expect to see a significant reduction in nutrient loading to the Mississippi River and over time, the Gulf of Mexico (McIsaac et al. 2002; Chapters 7 and 15, this volume).

In Scenario 1, agricultural commodity production is assumed to be the dominant objective of land management. Policy encourages cultivation of all highly productive land and assumes an adequate supply of fossil fuels or available equivalents, chemicals, and technology to a degree similar to the baseline. In Scenario 2, agricultural enterprises change in response to policy that enforces clear, measurable, farm-scale performance standards for water quality, supports practices to clean and detain storm water, and supports agricultural enterprises that help to achieve those standards, including rotational grazing in integrated crop and livestock enterprises. In Scenario 3, policy supports increased abundance and diversity of native plants and wildlife in the context of agricultural landscapes. Public investment creates a comprehensive system of indigenous species reserves of at least 640 acres (260 ha) connected by a network of wide habitat corridors and innovative agricultural practices that also buffer streams and protect water quality (Tables 4-1 and 4-2, this volume).

## Alternative Future Scenario Assessments

For all three scenarios, the alternative futures were compared to one another and to the baseline condition of the study watersheds using GIS-based models and digital simulations as described in Chapter 4 (see Color Figures 4–18). Changes in land use and management in the alternative futures were evaluated for their impacts on water quality (discharge, export of total suspended sediment [TSS], and export of nitrate); estimated financial profit; farmer preferences; and impacts on native plants and animals, as described in detail in Part 2 of this volume and elsewhere (Coiner et al. 2001; Vaché et al. 2002; Rustigian et al. 2003; Santelmann et al. 2004, 2005). Measurement and modeling methods used for assessments by the disciplinary teams are outlined briefly here.

To evaluate the potential profitability of the alternative futures, the Interactive Environmental Policy Integrated Climate (i_EPIC) model developed by Williams et al. (1988) was used (Chapter 5). Farmer preferences for the alternative futures were summarized and mapped using a spatially explicit method developed by Nassauer and Corry (Chapter 6). The Soil and Water Assessment Tool (SWAT) developed by Arnold et al. (1995) was used to evaluate scenarios for water quality response (Chapter 7). For evaluating risk to biodiversity, three approaches were used and compared. The first was a statistical estimate of change in habitat area, weighted by habitat quality (Chapters 8, 9, and 13). The second approach used spatially explicit population models (SEPMs) to estimate changes to biodiversity over time (Chapters 10 and 11). Finally, landscape pattern metrics (LPMs) were calculated (Chapter 12) and compared with the spatially explicit population model in Chapter 11.

Models such as those used in this study can only approximate the watershed-level response of various endpoints to changes in land use, but they allow us to rank alternatives with respect to their effectiveness in achieving the desired objectives (Starfield 1997). Long-term ecological research in agricultural systems is needed to study the watershed-level impacts of changes in land use and management, to validate the model results presented here, and to explore environmental responses to agricultural practices across scales.

Most efforts to develop landscape-level management strategies have focused on conservation planning and diversity maintenance based on general theoretical principles of habitat fragmentation; intensive single-species case studies; or static, habitat-based analyses (Pressey et al. 1993; Rabb and Sullivan 1995; Collinge 1996). Comparison of results from different models used in our project as well as results from other studies indicates that habitat loss, changes in landscape pattern, environmental variability, and interspecific interactions can all significantly influence the potential diversity of native species. This integrated assessment is a basis for further investigation of agricultural landscape futures and watershed-level field experiments to quantify effects of changes in agricultural land use and management. The array of approaches employed here helps to explore and quantify the magnitude and direction of each of these effects.

The economic endpoint for financial return to land (RTL, a measure of total watershed profit generated from agricultural operations; Chapter 5) used to evaluate the scenarios is very different from the social endpoint (farmer perceptions; Chapter 6). The financial endpoint attempts to precisely estimate RTL from crop yields and farm enterprise budgets, but the social endpoint employs multiple methods, using qualitative data to judge the validity of quantitative indicators of the acceptability of the landscape as a whole (landscape preference). The farmer preference study was intended to serve as an indicator of which scenarios and scenario components Iowa farmers perceive as good for the future of the people of Iowa, and to help identify innovations that farmers might find acceptable. Although no estimate of human preferences or attitudes is a fully reliable predictor of behavior (Arcury 1990), information about public perceptions can be useful for decisionmakers.

## Integration of Results: Multiple Endpoints on a Common Scale

To present results of multiple endpoints on a common scale for comparison, we expressed each endpoint in terms of percent change relative to baseline conditions (the difference between the future and present value of the endpoint divided by the present value) such that positive change reflected improvement relative to goals of the scenarios. We considered the following changes to be improvements: (1) greater profitability (RTL); (2) greater area in land use preferred by farmers as better for the people of Iowa; (3) increase in land cover associated with high native plant diversity or high suitability for native animal species; (4) increase in estimated potential abundance of native species; (5) decrease in export of nitrate and TSS; and (6) decrease in annual discharge. When more than one entity was being used to constitute the indicator (such as a set of plant or animal species for biodiversity), we used the median of the individual percent changes.

Each scenario implemented as a landscape future for Walnut Creek and Buck Creek was assessed for the ten indicator endpoints discussed in this section, for which percent change was calculated as described previously.

*Endpoints 1–3* represent water quality response: (1) annual discharge in $m^3$ $yr^{-1}$; (2) annual export of sediment in t $yr^{-1}$; and (3) annual export of nitrate-nitrogen in kg $yr^{-1}$. These three endpoints were estimated using the SWAT model as described in Chapter 7 (Vaché et al. 2002).

*Endpoint 4* represents profitability as measured by annual RTL, summed for watershed (U.S. dollars). This endpoint was estimated using i-EPIC, as described in Chapter 5 and Coiner et al. (2001). It does not include any income from federal programs, which would be in addition to profits shown here. Instead, RTL includes only income from marketing commodities produced, and it does not anticipate changes in commodity prices related to production of corn-based or cellulosic ethanol. In addition, this end-point does not include economic benefits or costs resulting from environmental and societal characteristics of each scenario.

*Endpoint 5* represents farmer preferences as a cumulative mean preference rating for land covers in each future, based on farmer rating of images related to each land cover. This endpoint was calculated from data gathered from farmer interviews (Chapter 6; Color Figures 44–45).

*Endpoints 6–10* represent biodiversity response, in which *Endpoint 6* is an index of native plant biodiversity (change in land cover area of habitat for all native plant species [$n = 932$]; see Chapter 8). *Endpoint 7* is an index of butterfly biodiversity (change in land cover area of habitat for all butterfly species [$n = 117$]; see Chapter 9). *Endpoint 8* is an index of native vertebrate biodiversity (change in land cover area of habitat for all bird, mammal, reptile, and amphibian species [$n = 239$]; see Chapter 13; Color Figures 47–49). *Endpoint 9* is an index of response for mammal species (change in land cover area of habitat for all mammal species [$n = 50$]; see Chapter 11). *Endpoint 10* is an index of response for four amphibian species (median percent change in mean number of breeding females in simulation year 100; see Chapter 10.) We summarize the individual endpoint assessments in the sections that follow.

## Water Quality

The results of water quality modeling (see Chapter 7) indicate that the changes in land use and management practices envisioned in Scenarios 2 and 3 could lead to substantial improvements in water quality. For example, reductions in annual stream discharge for Scenarios 2 and 3 were estimated to be 32 and 54 percent, respectively, for Buck Creek. For Walnut Creek, annual stream discharge reductions were estimated to be more than 50 percent for both scenarios (Endpoint 1; Figure 14-1a,b). We interpret the decrease in stream discharge as a positive feature of the scenarios, caused largely by a decrease in storm runoff that makes more water available for infiltration and groundwater recharge. Even Scenario 1 showed some improvement in discharge relative to the baseline landscape (18 percent for Walnut Creek and 11 percent for Buck Creek), primarily resulting from the comprehensive adoption of no-till practices in this scenario.

Export of TSS (Endpoint 2; Figure 14-1a,b) is similar in response to discharge, with Scenarios 2 and 3 showing reductions of 67 and 55 percent, respectively, in Walnut Creek, and Scenario 1 reducing sediment export by 11 percent relative to the baseline landscape. In Buck Creek, reductions in TSS for Scenarios 2 and 3 were 45 and 37 percent, respectively, with a reduction of 10 percent in Scenario 1 relative to the baseline landscape.

Nitrate export (Endpoint 3; Figure 14-1a,b) shows reductions of more than 50 percent in Scenarios 2 and 3 in both watersheds. For Scenario 1, the model estimated an 8- to 19-percent increase in nitrate export relative to the baseline landscape. For comparison, in the base year landscapes of Walnut Creek and Buck Creek watersheds, annual discharge estimated by SWAT was $9.5 \times 10^6$ and $14 \times 10^6$ m$^3$ yr$^{-1}$, respectively; annual export of nitrate was estimated at $63.4 \times 10^3$ and $54.4 \times 10^3$ kg yr$^{-1}$, respectively, and annual export of sediment was estimated to be 532 and 1,673 t yr$^{-1}$, respectively.

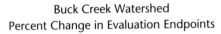

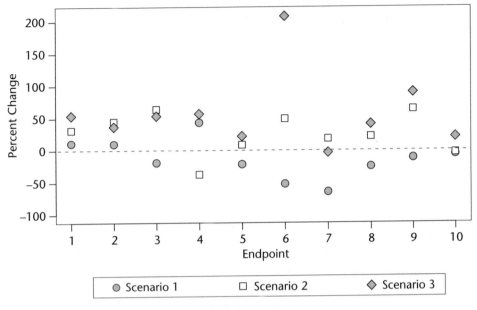

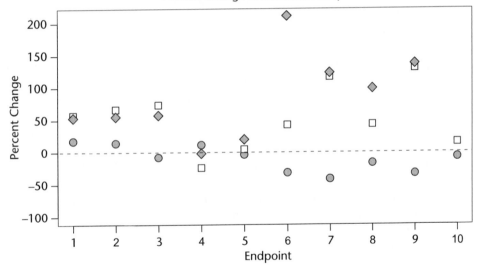

**Figure 14-1.** Results of the Ten Endpoints Modeled for Buck Creek (a) and Walnut Creek (b)

*Notes:* Results are presented as percent change relative to the baseline. Scenario 3 ranks consistently above Scenario 1 for all endpoints in Buck Creek, and for all but profitability in Walnut Creek. Scenario 2 ranks above Scenario 1 for all but economic profitability (Endpoint 4). Endpoints of plant and amphibian biodiversity (Endpoints 6 and 10) exhibit the greatest change and the greatest differences among scenarios, with Scenario 3 yielding the greatest improvement in biodiversity relative to the present (the change in amphibian biodiversity for Walnut Creek was 916%, off the regular scale of the figure).

## Profitability

Surprisingly, Scenario 3 ranked highest in terms of profitability in Buck Creek, and was nearly as profitable as Scenario 1 and the baseline landscape in Walnut Creek. For Scenario 3, RTL was almost unchanged compared with the baseline (see Table 14-1) in Walnut Creek and 30 to 40 percent higher in Buck Creek (Endpoint 4; Figure 14-1a,b). In Buck Creek, the relatively high RTL in Scenario 3 was realized partly because large areas changed from pasture or set-asides in the baseline landscape to perennial strip intercropping in Scenario 3. Perennial strip intercropping was estimated to be more profitable because the price of perennial prairie seed sold as a crop was estimated to be high relative to corn or beans (Color Figure 18). Although in Walnut Creek, Scenario 3 had less area in agricultural production compared with the baseline (partly because of the large bioreserves and wider stream buffers seen in Color Figures 5, 15, 42, and 43), Scenario 3 practices and enterprises were sufficiently profitable to yield RTL similar to that estimated for the baseline landscape (Chapter 5). Because the sale of corn or perennial cover for biofuels was not included in these RTL models, the potential profitability of biofuel enterprises is not part of these results. Either perennial covers such as those included in Scenario 3 (Color Figures 5 and 18) or use of corn (Color Figures 4, 11, and 16) for biofuels could change these results.

## Farmer Preferences

Farmers indicated an openness to changes in agricultural patterns and practices, as well as a high level of sophistication about ecological characteristics and environmental effects of agriculture (Chapter 6; Color Figures 4–27 and 45) In each watershed, farmers most preferred Scenario 3, followed by Scenario 2 and the baseline landscape. Scenario 1 was least preferred for both watersheds (Endpoint 5: Figure 14-1a,b). Several images of Scenario 2 also ranked quite high in farmer preference. When farmers indicated a preference for Scenario 2, they said that they valued the fact that it allowed more farmers to remain in farming. The extensive, pastured livestock operations and the corn–soybean–oat–alfalfa rotations of Scenario 2, though, were the least profitable agricultural enterprises modeled (Chapter 5) given agricultural commodity pricing in the RTL model. Scenarios 2 and 3 showed substantial water quality improvements (Chapter 7), but Scenario 3 performed substantially better in terms of profitability and overall farmer preference. Farmers' interview comments indicated that they typically expected Scenario 1 to be a *likely* future for Corn Belt agriculture, but they rated it far behind the other scenarios as being best for the future of the people of Iowa (Chapter 6).

## Biodiversity

As expected, Scenario 3 ranked highest on most endpoints for estimating impact to native biodiversity (Endpoints 6–9; Figure 14-1a,b). Only butterfly diversity in Buck Creek watershed ranked below baseline values for Scenario 3, because the conversion of pasture to strip intercropping in that scenario does not result in improved habitat for butterflies and because the upland bioreserves in Buck Creek watershed did not include wetland habitat (Chapter 9). Buck Creek watershed includes little area in wetland soils (and none in the parts of the watershed where reserves were located). As a result, very little wetland habitat was envisioned in the Scenario 3 future for Buck Creek. In Walnut Creek, where pasture and grassland make up less than 4 percent of the baseline landscape, and where bioreserves contained extensive prairie, wetland, and forest restorations, all native biodiversity endpoints (butterfly, plant, and vertebrate diversity and amphibian abundance), were substantially higher in Scenario 3 than for the baseline or Scenario 1 (Figure 14-1b).

For terrestrial vertebrate species (Endpoint 8; Figure 14-1b), biodiversity is almost double in Walnut Creek in Scenario 3 compared with a 17 percent decrease in Scenario 1. This overall improvement with Scenario 3 resulted not only from establishing reserves, a key element for amphibians (Chapter 10), but also from changing cropping and tillage practices. Converting to alternative perennial crops and using innovative agricultural practices, such as strip intercropping, modified large areas (e.g., the 83 percent of Walnut Creek in Color Figure 43 and the 45 percent of Buck Creek in Color Figure 42, which had been in row crops that provided little habitat in the baseline landscape in Color Figures 30 and 34).

## Integrated Assessment of Alternative Future Scenarios

Scenario 3 not only enhanced biodiversity more than any other scenario, as expected, it also was more profitable than the baseline landscape in Buck Creek (and nearly as profitable as the baseline landscape in Walnut Creek). In addition, it ranked highest in farmer preference and yielded substantial water quality improvements similar in magnitude to Scenario 2, which was designed to enhance water quality. If the changed cropping practices and land cover patterns seen in the study watersheds for Scenario 3 were applied throughout the Corn Belt, biodiversity would be enhanced throughout the Upper Mississippi Basin, and water quality would dramatically improve in the Corn Belt and, ultimately, in the Gulf of Mexico. Substantial improvements in water quality downstream from agricultural regions will require long-term, comprehensive change in agricultural practices (Becher et al. 2000; Schilling and Thompson 2000). Even though the conservation tillage and precision agriculture employed in all future scenarios contributed to improved water quality (Chapters 5 and 7), such practices alone are not sufficient to reduce the nutrients and sediment exported from cropland in corn and soybeans.

Water quality modeling results for these watersheds indicate that even with comprehensive adoption of conservation tillage under Scenario 1, nitrate export increases slightly and TSS decreases slightly. Under Scenarios 2 and 3, nearly threefold decreases in nitrate export and twofold decreases in sediment production are estimated (Chapter 7). Scenarios 2 and 3 are quite similar in predicted water quality (Endpoints 1–3; Figure 14-1a,b), and substantially better than either Scenario 1 or the baseline landscape. Although Scenario 1 specifies universal application throughout the watershed of conventional BMPs (universal use of no-till practices, 9.8- to 19.7- ft (3- to 6-m) buffer strips on permanent stream channels, filter strips, and precision farming targeting fertilizer applications to minimize nutrient additions), water quality benefits were minimal overall.

Of the three alternatives, Scenario 1 ranked lowest in terms of farmer preferences, water quality, and biodiversity endpoints. Only in profitability did it rank highest (for Walnut Creek, second for Buck Creek)—not a surprising finding because this scenario emphasized commodity production. What is surprising is that Scenario 3 actually achieved RTL nearly as high as did Scenario 1 for Walnut Creek and higher RTL than under Scenario 1 for Buck Creek. The high productivity of the strip intercropping employed in Scenario 3, combined with the high market price of the native seed produced in some of the strip intercropping in Scenario 3, helps to explain

**Table 14-1.** Results of the Ten Endpoints Modeled for Walnut Creek (a) and Buck Creek (b)

### (a) Walnut Creek Watershed

| Endpoint | Units | Baseline | Scenario 1 | Scenario 2 | Scenario 3 | Historical |
|---|---|---|---|---|---|---|
| | | *Water Quality* | | | | |
| 1. Annual discharge[a] | 103 ft3/yr (103 m3 yr –1) | 336,951 (9,541) | 276,464 (7,828) | 143,317 (4,058) | 156,658 (4,436) | 122,813 (3,478) |
| 2. Sediment export[a] | st/yr (t year –1) | 524 (532) | 448 (455) | 174 (177) | 234 (238) | 97 (99) |
| 3. Nitrate export[a] | lb/yr (kg yr–1) | 139,509 (63,413 ) | 150,286 (68,312 ) | 37,013 (16,824) | 59,442 (27,019) | 13,495 (6,134) |
| | | *Social and Economic* | | | | |
| 4. Watershed RTL[b] | Millions of Dollars | 1.550 | 1.731 | 1.176 | 1.523 | na |
| 5. Farmer rating[c] | Preference score | 2.09 | 2.02 | 2.19 | 2.50 | na |
| | | *Biodiversity* | | | | |
| 6. Native plants[d] | HC score | na | –31.9 | 42.3 | 211.5 | 262.4 |
| 7. Native butterflies[e] | HC score | na | –41.6 | 116.8 | 123.2 | 1189.9 |
| 8. Native vertebrates[f] | HC score | na | –16.8 | 42.9 | 98.9 | 33.8 |
| 9. Native mammals[g] | % change | na | –33.3 | 130.2 | 137.3 | na |
| 10. Native amphibians[h] | % change | na | –7.54 | 15.23 | 916.37 | na |

### (b) Buck Creek Watershed

| Endpoint | Units | Baseline | Scenario 1 | Scenario 2 | Scenario 3 | Historical |
|---|---|---|---|---|---|---|
| | | *Water Quality* | | | | |
| 1. Annual discharge[a] | 103 ft3/yr (103 m3 yr–1) | 495,129 (14,021) | 439,535 (12,446) | 338,013 (9,571) | 227,118 (6,431) | 263,988 (7,475) |
| 2. Sediment export[a] | st/yr (t year–1) | 1,647 (1,673) | 1,482 (1,506) | 907 (922) | 1,039 (1056) | 632 (642) |
| 3. Nitrate export[a] | lb/yr (kg yr–1) | 119,592 (54,360) | 142,426 (64,739) | 43,091 (19,587) | 55,636 (25,289) | 1,6245 (7,384) |
| | | *Social and Economic* | | | | |
| 4. Watershed RTL[b] | Millions of Dollars | 1.354 | 1.9401 | .847 | 2.121 | na |
| 5. Farmer rating[c] | Preference score | 2.15 | 1.68 | 2.33 | 2.62 | na |

**Table 14-1.** Results of the Ten Endpoints Modeled for Walnut Creek (a) and Buck Creek (b) (continued)

**(b) Buck Creek Watershed**

| Endpoint | Units | Baseline | Scenario 1 | Scenario 2 | Scenario 3 | Historical |
|---|---|---|---|---|---|---|
| | | | Biodiversity | | | |
| 6. Native plants[d] | HC score | na | -52.3 | 48.9 | 208 | 262.4 |
| 7. Native butterflies[e] | HC score | na | -64.6 | 18 | -3.5 | 164 |
| 8. Native vertebrates[f] | HC score | na | -25.1 | 21.4 | 40.6 | 40.6 |
| 9. Native mammals[g] | % change | na | -11.8 | 63.5 | 90 | na |
| 10. Native amphibians[h] | % change | na | -6.33 | -3.98 | 20.58 | na |

*Notes:* Scenario 3 ranks consistently above Scenario 1 for all endpoints in Buck Creek, and for all but profitability (as measured by watershed return to land) in Walnut Creek. Scenario 2 ranks above Scenario 1 for all but economic profitability. Biodiversity exhibits the greatest change among scenarios, with Scenario 3 yielding the greatest improvement relative to the baseline. HC stands for Habitat Change score. The formula for calculating the habitat change score, $HC_j$, for a specific group of species for a landscape is described in Chapter 13.

[a] Vaché et al. 2002

[b] Coiner et al. 2001

[c] Nassauer et al. Chapter 6, this volume

[d] Santelmann et al. Chapter 8, this volume

[e] Debinski et al. Chapter 9, this volume

[f] Santelmann et al. 2005

[g] Clark et al. Chapter 11, this volume

[h] Rustigian et al. 2003

this profitability. Proportional gains in profitability, even for Scenario 1, were very modest compared with proportional gains that could be achieved for other endpoints. Depending on market forces and policy, it may be possible to substantially improve environmental and societal benefits with little effect on baseline profitability.

The water quality and biodiversity model results support the perceptions of farmers in the region (Chapter 6), who identified Scenario 3 as best for the future of the people of Iowa, even though Scenario 3 is characterized by novel, challenging land management practices. Farmers identified Scenario 1, which employs familiar land management practices to achieve an intensification of agricultural production, as least good for the future of the people of Iowa (Color Figures 7, 11, 13, and 15).

Farmer perceptions may be influenced by some of the implicit scenario costs and benefits that were not incorporated in the profitability endpoints, which measure only RTL from the agricultural enterprise. The profitability endpoints did not estimate payments from federal programs that could encourage adoption of the practices in the scenarios, and did not estimate the value of quality of life or ecosystem services. For example, RTL does not include the economic benefit of improved water quality or enhanced native biodiversity or benefits at the national and international scale. Examples of these benefits include increased soil organic carbon (Robinson et al. 1996) to mitigate global climate change and reduced nitrate export from agricultural watersheds to coastal waters such as the Gulf of Mexico (Chapters 2 and 15). RTL does not include economic costs of environmental disasters resulting from flooding along the Mississippi River or from episodic failure of manure treatment facilities that might be associated with confined feeding operations. Incorporation of these costs would tend to decrease the profitability of Scenario 1 and enhance the overall profitability of Scenarios 2 and 3. Broad-scale economic analyses of various approaches to reducing nutrient overenrichment (McIsaac et al. 2002) indicate that combining multiple methods (for example, changes in farming practices coupled with wetland restorations such as those described in Chapter 15) are likely to be more economically efficient than those focusing on a single strategy (e.g., reducing fertilizer applications).

## Policy Implications

This integrated assessment allows us to generate and evaluate a broad range of innovative alternative futures and to identify those that are desirable for multiple endpoints. Policy, technology, and research can work in concert to realize the promising aspects of those options. Our results indicate that

1. If agricultural production continues to intensify, further degradation in water quality and loss of native biodiversity in agricultural regions will be a likely consequence.
2. Substantial improvements in water quality and native biodiversity could be achieved by implementing comprehensive changes in agricultural land use and management practices, assuming that public support and agricultural policy place high priority on clean water and restoration of habitat for native species.
3. Alternative future scenarios can be designed and implemented so that they are comparable to baseline farming operations in profitability and, at the same time, yield greatly enhanced environmental benefits.

**Table 14-2.** Summary of Effects of Policy Scenarios for Alternative Futures in Two Iowa Watersheds

|  | *Present* | *Scenario 1* | *Scenario 2* | *Scenario 3* |
|---|---|---|---|---|
| **Policy emphasis** | Produce commodities | Produce commodities and comprehensively apply BMPs | Improve water quality and hydrologic regimes while producing commodities | Improve biodiversity while improving water quality and producing commodities |
| **Water quality effect** | Baseline | Some improvement, but increased nitrate export | Substantial improvement | Substantial improvement |
| **Profit as RTL** | Baseline | Increase | Decrease | Increase |
| **Farmer preference** | Baseline | Decrease | Substantial increase | Substantial increase |
| **Biodiversity** | Baseline | Decrease | Substantial increase | Substantial increase |
| **Overall effect** | Baseline | Environmental and societal effects worsen overall, but RTL increases somewhat in some watersheds | Environmental and societal effects improve, but RTL decreases | Environmental and societal effects improve, and RTL increases somewhat in some watersheds |

4. Farmers are likely to recognize environmentally beneficial alternative scenarios as highly desirable and preferable to baseline agricultural practices as long as policy holds farm income effects neutral.

5. There is room for substantial improvement relative to baseline landscapes for all endpoints used to evaluate landscape response. Without sacrificing profitability, we can improve considerably on the present to achieve future societal and environmental benefits from Corn Belt agricultural landscapes.

Importantly, the studies that contribute to this integrated assessment provide a basis for policymakers to consider not only the particular futures described in Chapter 4, but also the potential effects of separate components of the futures (such as wetlands, riparian buffers, perennial cropping practices, woodlands, and other alternative cropping and management practices). There are numerous other ways that policymakers could choose to combine these components to drive changes in the ecological and socioeconomic endpoints toward nationwide conservation goals (Chapters 2 and 16).

This integrated assessment is a solid foundation for courageous innovation in agricultural conservation policy. It shows that practices and enterprises that were not part of conventional agriculture even a decade ago can be plausible for the future. We should not hesitate to imagine agricultural landscapes that could encourage Americans to look to the future with hope about the environment we will deliver to future generations.

## Acknowledgments

We thank the U.S. Environmental Protection Agency (U.S. EPA) STAR grants program (Water and Watersheds, grant #R-825335-01) for funding, those involved with the MASTER program for sharing with us data essential to this project, and all participants in the initial project workshop that laid the groundwork for the scenario designs. We extend special thanks to the Iowa farmers who participated in our interviews. Although the research cited here was funded by the U.S. EPA, the conclusions and opinions presented are solely those of the authors and are not necessarily the views of the agency.

*Chapter 15*

# Improving Water Quality from the Corn Belt to the Gulf

Donald Scavia, William J. Mitsch, and Otto C. Doering

In 2001, a task force of federal, state, and tribal representatives—established under the Harmful Algal Bloom and Hypoxia Research and Control Act of 1998—set a goal of reducing the size of the dead zone (the hypoxic region in the northern Gulf of Mexico) from its current size of more than 4.9 million acres (20,000 km²) to below 1.2 million acres (5,000 km²) (U.S. EPA 2001). In its action plan, the task force recognized that to reach that goal, nitrogen loads to the Gulf would have to be reduced by approximately 30 percent. Subsequent modeling analyses (Scavia et al. 2003, 2004; Justic et al. 2003a,b, 2005) indicate that, considering interannual variability in ocean conditions and potential long-term climate change, nitrogen loads should be reduced even further—in the range of 40 to 50 percent—to reach the task force goal.

Analysis of the causes and consequences of hypoxia in the northern Gulf of Mexico (CENR 2000; Rabalais et al. 2002) concluded that reducing the amount of nitrate-nitrogen (nitrate-N) reaching the Gulf can be accomplished by modifying agricultural practices; constructing and restoring riparian zones and wetlands; controlling urban and suburban nonpoint sources as well as atmospheric sources; and performing tertiary wastewater treatment at point sources (Mitsch et al. 1999, 2001). Significant work remains to be done, however, to determine where each of these measures would be most effectively targeted within the massive Mississippi River Basin (MRB). Goolsby et al. (1999) offer some insight into this question by evaluating the geographic distribution of major sources of nitrogen within the basin. Doering et al. (1999), Doering (2002), and Ribaudo et al. (2001) give further insight into the cost-effectiveness of the various approaches, comparing landscape patterns and practices similar to those included in the Iowa landscape futures described earlier in this book. In this chapter, we review some of those recommendations and assessments with the goal of providing a useful context for potential implementation at the local scales outlined in Part 2 and in the concluding chapter of this book.

Based on a formal review of the possible nitrogen reduction techniques listed in Table 15-1, Mitsch et al. (1999, 2001) concluded that a reduction of about 40 percent of nitrogen loading to the Gulf of Mexico is possible through a combination of several approaches:

- Changing farm practices by reducing the use of nitrogen fertilizer, managing manure nitrogen more carefully, and applying an array of best management practices (BMPs) on farms
- Creating and restoring wetlands, riparian forests, and buffer zones to intercept laterally moving groundwater and surface water from farmlands, especially in areas with artificial subsurface drainage and high concentrations of nitrates
- Managing existing and planned river diversions in the Mississippi River delta, particularly by intercepting large fluxes of nitrogen associated with floods
- Installing tertiary treatment systems for removing nitrogen from wastewater sources in the MRB.

## Means for Reducing Nitrogen Loads

Minimizing nitrogen loss (N-loss) from the land is the first line of defense for reducing nutrient loading to surface waters, groundwaters, and the Gulf of Mexico. This can be accomplished by changing farming practices to better manage farm nutrients. Applying fertilizer nitrogen at the correct rate—avoiding application at "insurance" levels (those beyond the minimum necessary for average production conditions)—and at the optimal time will substantially decrease nitrate losses. Knowing the nutrient content and application rate of manure, spreading it uniformly, and using it in the right season would all lead to better management of (and confidence in) manure nitrogen as a nutrient source. Giving appropriate nitrogen credits to previous legume crops and animal manure applications would also reduce excessive application. If nitrogen were better managed on farms throughout the MRB in these ways, an on-site reduction in nitrogen sources to streams and rivers via subsurface drainage of about 0.9–1.4 MMt/yr could be realized (Mitsch et al. 2001). One technology for achieving this reduction in excess nitrogen is the precision agriculture included in all the future scenarios for Iowa watersheds described in this book. Proper manure management is a key feature of Scenario 2, which features rotational grazing and requires that livestock be fenced from stream corridors.

Alternative cropping systems, in which perennial crops are planted in lieu of row crops in some areas, could also greatly reduce nitrate losses. Perennial crops such as the alfalfa and/or grass–alfalfa mixes that are widely used in Scenario 2 have been noted for their longer periods of nutrient and water uptake, which can optimize the cycling of nitrogen in the soil. If used as part of an alternate cropping system, these perennials could decrease nitrate loss by almost 90 percent in areas where those substitutions are made. Another perennial crop option could be the native herbaceous plants included in Scenario 3, in which native prairie grasses are grown as a crop and could be harvested for seed or as a feedstock for biofuel production. If changes of this kind were made to just 10 percent of the corn–soybean farms in the MRB, an estimated decrease in loading of 0.5 MMt/yr to streams and rivers could result (Mitsch et al. 2001).

Drainage tile lines that are required for growing corn and soybeans on former wetlands allow water to drain off-site quickly and effectively, but they also reduce potential infiltration and nutrient uptake. Even increasing tile spacing to greater than 49.21 feet (15 m) could continue to permit adequate drainage while limiting nitrate loss through subsurface drainage. Under Scenario 3, the perennial prairie strips in the strip

**Table 15-1.** Recommended Approaches for Reducing Significant Amounts of Nitrogen Loading to Streams and Rivers in the MRB and the Gulf of Mexico

| *Approach* | *Potential nitrogen reduction[a]* $2.2 \times 10^6$ *lbs./yr ($10^3$ Mt/yr)* |
|---|---|
| *Changing farm practices* | |
| Reduce insurance rates of nitrogen fertilizer application, distribute manure properly, apply appropriate credits for previous crop legumes and manure, and apply improved soil nitrogen testing methods | 900–1400 |
| Substitute perennial crops substituted for 10 percent of the current corn–soybean area | 500 |
| Improve management of animal manure in livestock-producing areas | 500 |
| Leave minimum spacing of 49.21feet (15 m) between farm drainage tiles | unknown |
| *Creating and restoring wetlands and riparian buffers* | |
| Create or restore 5–13 million acres (21,000–53,000 km²) of wetlands in the MRB (0.7 to 1.8 percent of the basin) | 300–800 |
| Restore 19–48 million acres (78,000–200,000 km²) of riparian bottomland hardwood forest (2.7 to 6.6 percent of the basin) | 300–800 |
| *Reducing point sources* | |
| Perform tertiary treatment of domestic wastewater | 20 |
| *Flood control in the Mississippi* | |
| Construct river diversions in the delta | 50–100 |

*Source*: Mitsch et al. (2001). Originally published in *Bioscience* 51(5): 382.
[a] Estimated on-site source reductions do not translate to equivalent reductions in Gulf of Mexico nitrogen loading, because only about 8 percent of nitrogen sources reach the lower Mississippi River.

intercropping practice could also serve as less-well-drained collection areas between tile drains to avoid crop loss in wet years.

Wetlands, which are widely understood as effective nitrogen sinks, are a second line of defense for reducing the nutrient load in the region. In the Corn Belt, presettlement land cover was dominated by wetlands (Chapters 2, 3, and 4); after drainage, these areas became highly productive farmland. Wetlands and riparian zones, though, can buffer the movement of nutrients from agricultural land to streams and rivers. Constructed wetlands, including off-channel floodplain storage, can be designed to enhance nitrogen load reduction through denitrification and nitrogen uptake by plants. Plant uptake is important only if nitrogen is stored in the soil or biomass for a long time or plants are harvested and removed.

Studies cited in Mitsch et al. (2001) suggest nitrogen reduction rates of about 4 g N/m²/yr for riparian forests and 10–20 g N/m²/yr for restored or created wetlands. Based on those rates, approximately 5–13 million acres (21,000–52,000 km²) of created and restored wetlands in the MRB (0.7 to 1.8 percent of the basin) would result in reductions of $660 \times 10^3 - 170 \times 10^7$ lbs./yr ($300–800 \times 10^3$ Mt/yr) of N loading to the Gulf. Use of riparian zones instead of wetlands would require 20–49 million acres (78,000–200,000 km²) to effect the same nutrient load reduction (Mitsch et al. 1999, 2001).

Beyond the Corn Belt, diversion of more river reaches within the deltaic plain in Louisiana could also help distribute more nutrients into areas that are natural sinks—marshes, swamps, and shallow open coastal areas in the delta—before they reach the Gulf. Although the overall N-loss from this option is relatively small compared to other options (Table 15-1), restoring delta wetlands could have important ecological and flood protection benefits. For example, at a controlled diversion at Caernarvon, Louisiana, nitrate removal was greater than 90 percent at a loading rate of about 10 g $N/m^2/yr^1$ (Lane et al. 1999). In the Louisiana deltaic plain, the Atchafalaya River carries about one-third of the Mississippi River discharges into a shallow inshore area, and about 50 percent of the nitrate is removed before the river plume reaches the stratified nearshore region. Assuming a retention rate of 10 g $N/m^2/yr^1$ (Lane et al. 1999), a reduction of $110 \times 10^3 - 220 \times 10^6$ lbs./yr (50–100 $\times 10^3$ Mt/yr) of nitrogen (Table 15-1) could be achieved by diverting Mississippi River water into the delta. For example, removing $110 \times 10^6$ lbs./yr (50,000 Mt/yr) would require diversion of about 13 percent of total river flow of a delta area of about 1.2 million (5,000 $km^2$).

The proportion of the overall nitrogen load to the Gulf from sewage treatment plants is relatively small under current conditions. Sewage point sources have been addressed by past policy implementation of the Clean Water Act of 1972. To achieve further gains, wastewater wetlands and other ecotechnologies could be used to treat water from municipal wastewater treatment and other point sources. But applying tertiary treatment systems such as constructed wetlands and nitrification–denitrification basins to sewage treatment plants to reduce the nitrogen discharged from remaining point sources by 50 percent would reduce the overall nitrogen load to the Gulf by only a small percentage (Table 15-1). In comparison, managing nonpoint sources from agricultural land would have a substantially greater effect on reducing nitrogen load to the Gulf.

## Geographic Targeting

The individual methods we have outlined will not be equally effective everywhere in reducing nitrogen loads to the Gulf. The challenge is to select and adopt the most appropriate method for local landscape conditions, anticipating the multiple ecological and societal effects of each method, and weighing the relative efficiency of each method for achieving MRB nitrogen reduction goals. Differing landscape conditions and varying agriculture and forestry enterprises characterize different regions within the MRB, and geographic targeting of different nitrogen reduction methods could fit these regions. Goolsby et al. (1999, 2001) evaluated the geographic distribution of major sources of nitrogen within the basin, and this information can help point to regions within the MRB where nitrogen reduction is needed most.

Given the assumption that there is little N-loss in large rivers, Goolsby et al. (1999) estimated that the Middle Mississippi Basin (i.e., the Corn Belt), with only 8.5 percent of the MRB drainage area, contributes about 33 percent of the nitrate discharging to the Gulf and the largest amount of nitrate and total nitrogen per unit area (Color Figures 50 and 51). The respective nitrate and total nitrogen yields (i.e., N-loss from a watershed divided by the area of the watershed) from this basin are 10.26 and 15.08 lb/acre/yr (1,150 and 1,690 $kg/km^2/yr$)—nearly 90 percent higher than the Ohio Basin and more than 100 percent higher than the Upper Mississippi Basin. About 56 percent of the nitrate and 54 percent of the total nitrogen comes from the Upper and Middle Mississippi River basins combined (i.e., the MRB above the Ohio River Basin). In comparison, the Ohio River Basin, on average, contributes about 34 percent of the nitrate and 32 percent of the total nitrogen discharged by the MRB to the Gulf. The Missouri Basin contributes about 15 percent of the total nitrogen. Nitrogen yields in the western half of the MRB are relatively low—2.86 lb/acre/yr (320 kg/km$^2$/yr) or less, and less than the entire MRB average of 4.36 lb/acre/yr (489 kg/km$^2$/yr). This can be attributed largely to the drier climate, lower runoff, and different land uses in the western region. The Lower Mississippi Basin contributes less than 8 percent of the nitrogen, and the combined Arkansas, Red, and Ouachita basins contribute less than 8 percent.

Goolsby et al. (1999, 2001) also provided yield estimates for many smaller watersheds within the major river basins of the MRB. They showed that the spatial distribution of total nitrogen and nitrate-N yield is highly varied among smaller watersheds (Color Figure 52). The highest average annual total nitrogen yields range from 8.92-26.77 lb/acre/yr (1,000 to more than 3,000 kg/km$^2$/yr) and occur in a band extending across the Corn Belt from southwestern Minnesota into Iowa, Illinois, Indiana, and Ohio. Annual average yields ranging from 1,800 to more than 27.66 lb/acre/yr (3,100 kg/km$^2$/yr) occur in the upper Illinois River Basin; the Great Miami Basin in Ohio; and the Des Moines and Skunk River basins in Iowa. The Skunk River Basin contains both of the second-order watershed study areas (Buck Creek and Walnut Creek) described in Part 2 of this book.

In 1993, for example, the nitrate discharged from these four basins alone was equivalent to more than 37 percent of the nitrate discharged to the Gulf. During years with high precipitation, the total nitrogen yield from these four river basins can range from 26.77 to 62.46 lb/acre/yr (3,000 to more than 7,000 kg/km$^2$/yr). Their discharge alone can account for as much as 21 percent of the total nitrogen discharge from the MRB during average years and more than 30 percent during flood years (such as 1993).

The Minnesota River Basin, Rock River and the Lower Illinois River in Illinois, the Grand River in Missouri, the Wabash River in Indiana, and Muskingham and Scioto rivers in Ohio have lower nitrogen yields of 8.92-16.06 lb/acre/yr (1,000–1,800 kg/km$^2$/yr). Other basins adjacent to these may have similar nitrogen yields. Nitrogen yields were generally 4.46-8.92 lb/acre/yr (500–1,000 kg/km$^2$/yr) in basins south of the Ohio River and generally less than 4.46 lb/acre/yr (500 kg/km$^2$/yr) in the Missouri, Arkansas, and Lower Mississippi basins. Many of the drier basins in the western part of the MRB had nitrogen yields of less than 0.89 lb/acre/yr (100 kg/km$^2$/yr).

This assessment suggests that efforts to reduce nitrogen loads to surface waters and the Gulf of Mexico would be most effective in the regions with the highest current nitrogen yields, including the second-order watersheds examined in this volume. Both altered farming practices and other means for reducing nitrogen loads could be targeted to locales where they will be most effective. For example, if 7 percent of the Illinois River Basin watershed were converted to wetlands, as much as 50 percent of the 317 x 10$^6$ lbs. (144 × 10$^3$ Mt) of nitrogen flow from that watershed to the MRB could be eliminated. In contrast, the James River Basin in South Dakota contributes only 2.6 × 10$^6$ lbs./yr (1.2 × 10$^3$ Mt/yr) of nitrogen to the basin. If all of these were controlled, the reduction of the nitrogen load to the Gulf of Mexico would still be insignificant. Clearly, restoration and creation of wetlands must be strategic; that is, wetlands should be located where agricultural sources of nitrogen and subsurface drainage are the greatest.

## Cost-Effectiveness and Economic Flexibility

As shown in Table 15-1, Mitsch et al. (1999, 2001) compare various options for keeping nitrogen on the land and removing it from surface and groundwaters. Goolsby et al. (1999, 2001) provide a geography of the largest sources, where options on that menu could be most effective. Doering et al. (1999), Doering (2002), and Ribaudo et al. (2001) give insight into the cost-effectiveness of the options and demonstrate that there is enough cost equivalency among them to allow them to be targeted where they would do the most good.

Doering et al. (1999) used the U.S. Agricultural Sector Mathematical Programming (USMP) regional agricultural model (House et al. 1999) to combine agricultural-sector adjustment costs and intrabasin benefits to calculate a net per-unit cost for nitrogen reduction. The USMP model is driven by production-related modeled decisions by farmers who maximize profits within specified constraints; it is a good tool for assessing modifications of production and practices to maximize profits under nitrogen restrictions or efforts to reduce N-loss.

The USMP model is a spatial and market equilibrium model designed for general-purpose economic and policy analysis of the U.S. agricultural sector. It predicts how technology or changes in farm resource, environmental, or trade policy commodity demand will affect regional supply of crops and livestock; commodity prices and demand; use of production inputs; farm income; government expenditures; participation in farm programs; and environmental indicators (such as erosion, nutrient loadings, and greenhouse gases). Other inputs, such as fertilizer and seed, assume fixed national level prices; land is modeled as crop and pasture classes; and labor is specified by family and hired types. Production systems are differentiated according to tillage, multiyear crop rotation, dryland/irrigation, government program participation, and other characteristics. Various environmental impacts from production, such as N-loss to leaching and runoff and soil erosion, are computed using the Erosion Productivity Impact Calculator (EPIC) biophysical model (Williams et al. 1988), the same model used to assess the market return to land (RTL) for the two Iowa study watersheds (Chapter 5). Wetland creation analysis was accomplished by first screening acreage and production affected by wetland restoration, followed by analyzing the impact of wetland restoration using USMP. Decisions were driven by least-cost considerations. Restoration costs included both the costs of permanent easements (the opportunity cost of removing productive cropland from production) and restoration costs. For further detail on the modeling and procedures see Ribaudo et al. (2001) and Doering et al. (1999).

Doering et al. (1999) compared different combinations of strategies for nitrogen reduction as applied to the entire MRB (Table 15-2). The estimated net costs vary considerably across the different combined options and for different load reduction goals. When other environmental and societal benefits as measured by EPIC, in addition to nitrogen reduction, are included in net per-unit cost, the cost of wetland strategies is reduced significantly, actually resulting in net benefits for the 1-million-acre option.

A critical reason that nitrogen reduction could be difficult is its extremely high value for agricultural production. Under the right circumstances, a pound of additional nitrogen can yield a return of tenfold its cost in corn production. This gives an extremely strong economic incentive for applying insurance levels of nitrogen, and helps explain why only a politically unacceptable 500 percent tax on fertilizer would be required to realize N-loss reductions equivalent to other scenarios. The buffer strategy is the least desirable because of the high opportunity costs of retiring land, low nitrogen filtering relative to wetlands, and small environmental benefits.

**Table 15-2.** Cost and Benefit Comparison of Strategies for Reducing Nitrogen Loading to Surface Waters and the Gulf of Mexico

| Strategy | Social welfare (million US$) | Erosion benefits[a] (million US$) | Wetland benefits[b] (million US$) | Net social benefits[b] (million US$) | N-loss reduction (1,000 Mt[c]) | Unit cost ($/Mt[c]) |
|---|---|---|---|---|---|---|
| **N-loss reduction (%)** | | | | | | |
| 20 | −831 | 78 | na[d] | −753 | 941 | 800 |
| 30 | −2,677 | 136 | na | −2,541 | 1,412 | 1,800 |
| 40 | −6,343 | 225 | na | −6,118 | 1,882 | 3,251 |
| 50 | −12,239 | 300 | na | −11,939 | 2,352 | 5,076 |
| 60 | −21,109 | 302 | na | −20,807 | 2,822 | 7,373 |
| **Fertilizer reduction (%)** | | | | | | |
| 20 | −347 | 12 | na | −335 | 503 | 666 |
| 45 | −2,922 | 39 | na | −2,883 | 1,027 | 2,807 |
| 500 percent fertilizer tax | −14,932 | 39 | na | −14,893 | 1,027 | 14,501 |
| **Wetland acres (hectares)** | | | | | | |
| 1 million (404,685.64) | −406 | 3 | 550 | 147 | 67 | −2,192 |
| 5 million (2,022,926.40) | −3,115 | 15 | 2,751 | −349 | 350 | 997 |
| 10 million (4,046,856.42) | −7,525 | 28 | 5,502 | −1,995 | 713 | 2,813 |
| 18 million (7,284,341.56) | −15,506 | 56 | 9,904 | −5,546 | 1,300 | 4,266 |
| **Buffer** | −18,014 | 62 | | −17,952 | 692 | 25,942 |
| *20 percent fertilizer reduction and 5-million-acre (2,022,926.40) wetland* | −4,854 | 26 | 2,751 | −2,077 | 882 | 2,354 |

[a] Social welfare for wetland strategies includes changes in consumer and producer surpluses, plus wetland restoration costs.
[b] Net social benefits include social welfare, erosion benefits, and wetland benefits.
[c] 1 Metric Ton = 2204.6 lbs.
[d] "na" indicates that this benefit was not measured in this study.
*Source:* Doering et al. (1999).

It is worth noting that constraining N-loss from the farm in the USMP (the N-loss strategy in Table 15-2) is more efficient than restricting fertilizer nitrogen use. In Part 2 of this book, Scenarios 2 and 3 both describe many practices and land use patterns that could constrain N-loss from the farm. To get the equivalent N-loss reduction ($2 \times 10^9$ lbs. (941,000 Mt) from the N-loss strategy; $2.2 \times 10^9$ lbs. (1,027,000 Mt) from fertilizer restriction strategy), the fertilizer restriction strategy costs about 3.5 times more than the N-loss strategy. This is because farmers in the USMP are allowed to modify their actions to reach the target in the most economically favorable way. Although theoretically more efficient, measuring N-loss from farms is difficult, complicating implementation and verification as well. Knowing the measurable outcomes of implementing practices such as those described under Scenarios 2 and 3 would be important for understanding the consequences of their widespread application.

Doering et al. (1999) also analyzed a mixed-policy strategy (Table 15-2) that restores 5 million acres of wetlands and applies a 20 percent nitrogen fertilizer restriction that they believed is the most practicable, cost-effective approach for meeting a 20 percent N-loss goal. In terms of cost-effectiveness, the 45 percent nitrogen fertilizer constraint comes close to the mixed-policy strategy, and if a lower N-filtering capacity is assumed for wetlands, it would be the most cost-effective. The mixed policy has a smaller impact on prices, resulting in smaller adjustments inside and outside the basin. The mixed policy resulted in a 1.6 percent increase in cropland outside the basin. The fertilizer reduction resulted in a 5.6 percent increase in cropland inside the basin.

Ribaudo et al. (2001) and Doering (2002) compared the policy cost implications of some of these strategies. In their application of the USMP (Figure 15-1), they compared net per-unit costs for wetland and fertilizer restriction, but in their analysis, wetland restoration was targeted proportional to total nitrogen yield by hydrologic unit as estimated by Smith et al. (2000). In other words, wetlands were restored first in those areas with the greatest N-loss. They assumed that all cropland on hydric soils in the National Resource Inventory (NRI) is eligible for enrollment except land converted in violation of the Swampbuster provisions of the 1985 Farm Bill (the Food Security Act of 1985). In addition, the amount of land that could be converted in a given, similar local area was limited to the 25 percent allowed for the Conservation Reserve Program (CRP). This resulted in a wider geographic distribution of wetland acreage, which also increased costs. Their analysis suggests that fertilizer restriction is more cost-effective than wetland restoration up to about a 25 percent reduction in N-loss ($2.7 \times 10^6$ lbs. [1,250 Mt]).

Both sets of studies showed that for N-loss reductions ranging from 25 to 35 percent, costs associated with fertilizer restriction and for wetland restoration are comparable (Figure 15-1), and not appreciably different from costs associated with the N-loss strategies (Table 15-2). This suggests that there is considerable cost equivalency in applying the methods outlined in Mitsch et al. (1999, 2001). This would facilitate the geographical targeting of approaches that follows from Goolsby et al. (1999, 2001).

## Different Approaches for Different Watersheds

To improve water quality from the Corn Belt to the Gulf, different agricultural watersheds could and probably should adopt different approaches. Changed agricultural practices and land use patterns in the heart of the Corn Belt could dramatically effect nitrogen loading to surface waters and the Gulf because nitrogen loading rates are very high in those watersheds now (Color Figures 50 and 51). Those changes could range

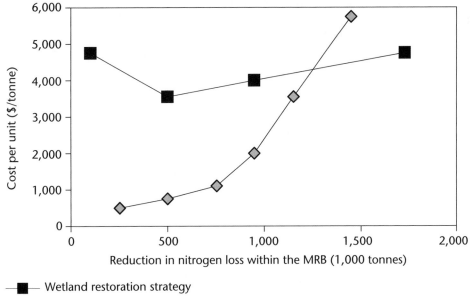

**Figure 15-1.**   Cost Comparison of Two Strategies for Reducing Nitrogen Loading: Fertilizer Reduction and Wetland Restoration

*Source:* Ribaudo et al. (2001). Reprinted from *Ecological Economics,* Volume 37 (2), Ribaudo, et al., 183–198, May 2001, with permission from Elsevier.

across the strategies outlined in Table 15-2—from the reduced fertilizer application modeled for all three of the scenarios, to landscape patterns such as those in Scenarios 2 and 3 (which employ perennial cover to reduce N-loss), to wetland creation and restoration to detain nitrogen before it moves downstream. Chapter 4 details several approaches (under Scenarios 2 and 3) to locating created wetlands and restoring wetland habitats. These include detaining water at tile drain outlets, maintaining a necklace of upstream detention wetlands, and providing off-channel storage along second-order streams. Doering (2002) and Ribaudo et al. (1999) each concluded that wetland strategies and fertilizer reduction strategies are about equally cost-effective at the level of nitrogen reductions desired for the entire MRB. Doering's strategies (Table 15-2) show that landscape patterns and practices to prevent N-loss may be the most cost-effective of all the strategies. It will be more difficult, though, to measure water quality results attributable to changed practices to prevent N-loss on any given farm.

Both reduced fertilizer applications and prevention of N-loss can be achieved using many of the BMPs and precision agriculture technologies that farmers are using widely today. In addition, returning to some crop rotations that were more widely used in past mixed-grain and livestock enterprises, such as the alfalfa hay and rotational grazing practices included under Scenario 2 (Chapter 4), could be effective in some locales. In other areas, restoring aspects of wetland and riparian ecosystems that have nearly disappeared from the Corn Belt would be important. Some new practices, patterns, and enterprises may result in landscapes that look very different from past Corn Belt landscapes. Permanent bioreserves, such as those included in Scenario 3, could be strategically located for their effects in preventing N-loss to streams. The perennial strips in Scenario 3 might not only provide habitat and contribute to new seed or biofuel enterprises, they could detain stormwater and nitrogen as well.

Variability in nitrate loading across the subbasins of the entire MRB strongly suggests trends for future agricultural policy. Federal agricultural policy that more broadly applies the watershed approach of the Conservation Security Program (Chapters 3 and 16) and guides farmers to options appropriate to the landscape and enterprise characteristics of their own watersheds could effectively relieve the hypoxia in the Gulf of Mexico.

*Chapter 16*

# Agricultural Policy Choices

## Otto C. Doering, Catherine L. Kling, Joan Iverson Nassauer, and Donald Scavia

Now, even more than in the past, American agricultural policy must balance national interests in agricultural production with fundamental interests in environmental health. Rapidly emerging opportunities for biofuels production and pressing demands by America's global trading partners, along with increasing societal attention to both the taxpayer costs and environmental effects of American agricultural policy, create momentum for policy change. In this book, integrated assessments at two scales, the Mississippi River Basin (MRB; Chapter 15) and second-order Iowa agricultural watersheds within the MRB (Chapter 14), suggest that improved policies could alleviate hypoxia of the Gulf of Mexico, as well as a wide range of environmental and societal impacts of agriculture in the Corn Belt. In addition, federal agricultural support that aims to achieve environmental benefits complements world trade objectives (Chapter 3). In this chapter, we discuss how federal agricultural policy could respond to these demands and opportunities for change.

## Implications for Landscape Change: Integrated Assessments at Two Scales

Across the MRB, nutrients coming from agricultural land uses are the chief cause of Gulf hypoxia, and Corn Belt watersheds are the leading contributors to nitrogen loading of the Gulf (Chapter 15; Color Figure 52). Changed agricultural practices, enterprises, and landscape patterns could reduce nutrient losses and ameliorate a wide range of environmental problems, including downstream flooding, impoverished biodiversity, climate change, and disappearing rural communities and institutions (Day et al. 2005a; Chapter 3, this volume). Solutions to these related environmental and social problems require a long-term, basinwide approach that links changes in Corn Belt agricultural landscapes to local benefits (such as enhancing quality of life in the Corn Belt), as well as to benefits further downstream (such as reducing hypoxia and helping to protect floodplain cities; Day et al. 2005b). Results of the integrated assessments at both scales suggest that enormous environmental benefits can be gained by employing conservation successes of past and current federal agricultural policy, allowing those successes to be adapted to evolving goals, and also considering new ideas like those included in the Corn Belt futures.

Landscape changes have been recommended for the entire MRB (Chapter 15), but for these changes to be effectively implemented, they must be adopted by individual farmers at the local scale, like the Iowa study watersheds described in Part 2 of this book. Assessments at both scales are necessary for making informed real-world judgments about policy, and both scales help point the way to the ultimate test of agricultural policy: choices made by farm operators and producers. Particularly in light of the local watershed focus of the Conservation Security Program (CSP, described in Chapter 3), the local watershed scale of the Corn Belt futures provides a more realistic frame for policy analysis than either political boundaries or farmed parcels alone. How recommended changes fit into agricultural enterprises, landscape patterns, and resource characteristics; how they respond to market and technological opportunities; how they affect farmer costs and incomes; and how they affect local quality of life will all influence farmers' decisions to adopt these practices. Examining how specific landscapes could actually change at the local watershed scale helps farmers and policymakers get a sense of these decisive characteristics.

The Corn Belt watershed alternative futures that are detailed in this book are intended to be guiding examples of how the types of practices recommended for the entire MRB could change agricultural landscapes. These landscape changes could yield broad societal benefits and ecosystem services, such as enhanced water quality, reduced flooding, greater biodiversity, healthier rural communities, and more desirable landscapes—all while protecting farmers' livelihoods (Chapter 14). Furthermore, the integrated assessment for the entire MRB (Chapter 15) suggests that combinations of practices such as those in the alternative futures could effectively address hypoxia as well. Finally, interviews with Iowa farmers suggest that they recognize these innovative practices as best for the future of the people of Iowa, as long as policy allows farmers' incomes to be unaffected by the new practices (Chapter 6).

Neither of the specific Iowa study watersheds is "typical" of the MRB. But because very few (if any) watersheds across the nation represent an "average," the consistent results from the integrated assessment for the two very different Iowa watersheds—as well as results of subsequent, analogous studies of other Corn Belt watersheds at similar scales (Boody et al. 2005; Burkart et al. 2005)—suggest that the Corn Belt futures may offer ideas that could also be useful elsewhere.

The two Iowa study watersheds are very different in their soils and geomorphology. Their integrated assessment results, however, showed both watersheds realizing gains in water quality and farmers' perceptions under Scenario 2, and similarly improving water quality, flood reduction, overall biodiversity, and farmers' perceptions under Scenario 3 (Chapter 14). Results for the two watersheds were notably different only in their return to land (RTL) revenue: The flat landscape of the Walnut Creek watershed realized slightly greater returns under Scenario 1, while the rolling landscape of the Buck Creek watershed realized greater returns under Scenario 3 (Chapter 5). Because neither corn nor perennial crop biofuel markets were included in the RTL models, estimated financial return results could change as biomass energy technologies emerge (Schnepf 2006).

The alternative futures do provide insight on landscape changes (and related environmental effects) that could occur with stronger biofuel markets, however. For example, biofuel markets drove predictions that acres planted in corn in the United States would increase by 15 percent between 2006 and 2007 (USDA NASS 2007). Similarly, Scenario 1 assumes that all highly productive soil is cultivated, taking land out of perennial cover and increasing land in corn or soybeans by about 8% in flat Walnut Creek watershed and by about 38% in the more erodible rolling landscape of Buck Creek watershed (see Chapter 7, Table 7-1 and 7- 2 ). On the other hand, Scenarios 2 and 3 assume increases in perennial herbaceous cover to at least 25 percent of each watershed (see Chapter 7, Table 7-1 and 7- 2), changes that might be similar to what would occur if cellulosic ethanol production became technologically practical. Integrated assessments of the alternative futures, in one sense, suggest how different approaches to biofuel generation could have different environmental effects.

Based on the integrated assessment of the MRB, alternative policy strategies for reducing nitrogen loading to the Gulf of Mexico could be implemented using the types of practices and landscape patterns shown in the Corn Belt futures (summarized in Table 4-1 and Table 4-2 in Chapter 4, and modeled in Chapter 7). Key basinwide strategies that could be implemented in local Corn Belt watersheds are described in Chapter 15 and summarized here:

1. *Changes in farming practices,* including reduced fertilizer use, precision agriculture, and crop pattern and enterprise changes. Such modifications could reduce the nitrate-nitrogen that reaches streams and rivers in the MRB by 20 percent (Mitsch et al. 1999). Each of the Corn Belt futures incorporates changes to reduce nitrogen in surface and groundwater by employing precision agriculture (in all scenarios) and/or by targeting an increased area in perennial cover (in Scenarios 2 and 3). In Scenario 2, increased perennial cover is achieved by incentivizing mixed-grain/rotational-grazing livestock operations (Color Figure 17). In Scenario 3, perennial cover is augmented by working lands very differently—by employing perennial strip intercropping of working lands (with rotations of conventional crops such as corn and soybeans and permanent perennial crop strips, which could be marketed as native plant seed or biomass; Color Figures 18 and 29).

2. *Creation and restoration of wetlands and riparian forests* located between farmlands and adjacent ditches, streams, and rivers. As described in Chapter 15, restoring 1 to 5 million acres (4,000 to 20,000 km²) of wetlands in the MRB may be a cost-effective approach to reducing nitrogen loading to the Gulf, particularly when other benefits of wetlands are included. This approach is included in different forms in alternative Scenarios 2 and 3 (Part 2). Under Scenario 2, a network of detention wetlands, filter and infiltration strips, and tile drainage discharge ponds is woven into a matrix of cultivated land and pasture (Color Figure 12). In Scenario 3, large permanent bioreserves remove some tile drainage to restore wetlands, prairie, and riparian habitats; they also infiltrate and detain storm water and retain nitrogen (Color Figures 5 and 15). Scenarios 2 and 3 are similar in that both feature broad filter buffer strips for all perennial streams; buffer strips extend 49 feet (15 m) from the stream bank in Scenario 2 (Color Figures 8, 12, 40 and 41) and 98 feet (30 m) from the stream bank in Scenario 3 (Color Figures 5, 9, 42, and 43). Both scenarios also feature extensive areas of riparian forest within the buffer strip, with the forest extending into upland wooded pastures in Scenario 2 and riparian forest bioreserves being established in Scenario 3 (Color Figure 9). As the biodiversity assessments of the Corn Belt futures (Chapters 8–13) demonstrate, these connected riparian systems also create significantly improved habitats.

3. *Diversion of rivers into adjacent constructed and restored wetlands* all along the river courses. Buffer width is extended up to 196 feet (60 m) to accommodate extensive off-channel storage in stream floodplains in Scenario 2 (Color Figure 12). Under Scenario 3, restoring indigenous floodplain ecosystems and upstream wetland ecosystems within selected bioreserves (Color Figure 15) results in similar ecosystem services (Chapter 7).

The alternative futures also demonstrate how adopting time-tested soil and water conservation practices more comprehensively (in all scenarios); increasing perennial cover, especially on erodible soils and stream buffer strips (in Scenarios 2 and 3; Color Figures 40–43); and enhancing storm-water detention and encouraging infiltration upstream (in Scenario 2) could contribute to reducing nutrient loads across the entire MRB. Field-calibrated Soil and Water Assessment Tool (SWAT) models for the two Iowa study watersheds forecast that using combinations of these familiar and innovative farming practices could decrease median nitrate loads in Scenarios 2 and 3 anywhere from 57 to 70 percent over six years (Chapter 7).

## Context for Policy Change

Neither the integrated assessments of the MRB nor the Corn Belt futures determine what policy expenditures might be required to achieve the recommended changes. At both scales, though, continued federal involvement to inform and incentivize farmers' land management choices is assumed to be in the broad public interest. In addition, increasing the proportion of federal agricultural expenditures dedicated to achieving environmental benefits is strongly implied. To assess environmental benefits, the Corn Belt study explicitly described the alternative policy scenarios (Chapter 4, Table 4-1) as holding average farm net income constant under widely differing policy scenarios. Scenario 1 continues current practices and trends, leading to larger, fewer farms; threatened biodiversity; and dwindling rural communities (Color Figures 38 and 39). Scenario 2 includes a level of mixed-grain/livestock activity that has not been the trend in the recent past (Color Figures 40 and 41). The resulting pastoral landscape supports a plausible assumption that the broader public would appreciate the landscape as an amenity. Implied labor demands of mixed-grain/livestock enterprises support a plausible assumption that farm size, scale, and degree of specialization would increase less dramatically in Scenario 2 than in either Scenario 1 or 3, keeping more farmers on the land. Scenario 3 encompasses the greatest innovation in practices to enhance the environmental benefits of reduced nutrient loss and increased habitat (Color Figures 42 and 43). Its dramatically increased biodiversity and varied landscape pattern support the plausible assumption that the broader public would perceive the agricultural landscape as an amenity—with larger numbers of nonfarmers living in rural areas of the Corn Belt to stabilize small-town services and institutions. At the same time, Scenario 3 reasonably assumes that these societal benefits could be achieved even if farm structure continues to follow current trends toward larger, fewer farms.

Changing current trends in agricultural policy would clearly be difficult. A major attempt to do so on a national scale in the early 1980s may have slowed previous trends, but did not stop most of them. In 1981, Secretary of Agriculture Robert Bergland publicized existing trends and their probable consequences, and suggested alternative policies to mitigate trends such as concentration and changing structure of ownership and control (USDA 1981). Interest in these issues subsequently diminished, though, and no policies to change these trends were implemented. Today, subsidies to crop production remain a foundation of commodity production systems, even as environmental benefits, global trade mandates, domestic spending priorities, and the varied perceptions and concerns of the American public, rural communities, and farmers have begun to align themselves to point toward new policy directions. Some states, such as Iowa, Nebraska, and California, are working to realign agricultural policy priorities within their states. Directing policy trends to achieve broad societal benefits may require a convergence of interests and awareness to recognize the possibilities for a very different future.

## Policy Change and Environmental Benefits in Future Agricultural Policy

Several important assumptions suggest how environmental benefits might be achieved in future farm bills. Debate will include vigorous discussion of conservation and environmental programs. Changes in Congress could change the tenor of the debate about conservation and the environment, but sentiment may remain strong for retaining the basic structure of existing conservation programs, in which participation is voluntary, technical assistance is provided, and participants receive incentives. Hanrahan and Zinn (2005) suggest that the World Trade Organization (WTO) designation of allowable green box support for agriculture, which has been important to evolving policy in America's Organisation for Economic Co-operation and Development (OECD) partner nations in Europe and elsewhere (Chapter 3), could be a factor that moves American policy toward a goal of obtaining greater environmental benefits from public investment in farm programs. Even though agreement has not been reached on what is meant by green box payments or on the political/environmental objectives of existing conservation programs (USDA NRCS 2006), fitting farm programs into the framework of trade negotiations will undoubtedly be increasingly important in years to come.

Past trade negotiations and farm bill legislative processes, however, teach us that policy decisions about environmental benefits from agriculture are incontrovertibly bound to decisions about commodity programs. Conventional practices for commodity production have had indisputable environmental effects (Chapters 2 and 3), and past farm policy has invested in commodity programs to a greater degree than in agricultural conservation. In the past, intensity and single-mindedness have carried the day for commodity programs against concerns of consumers, environmentalists, and others (Schertz and Doering 1999).

By increasing returns to farmers, commodity programs can increase the profitability of more intensive production as well as production on more marginal lands. High prices for commodities, like the record corn prices logged in the spring of 2007, can do the same, and both have the capacity to increase environmental damage from soil erosion and nutrient loss. Although the magnitude of these effects is uncertain, a low cost of nitrogen relative to the price of corn does increase the economic amount of nitrogen farmers might apply and vice versa.

For example, from 2004 to 2006, corn prices were low, and near the end of the two-year period, fertilizer prices increased tremendously, dampening any increase in nitrogen use. This changed starting in September 2006 when the projections of demand for corn for ethanol production resulted in an increase in the price of corn of over 50 percent during harvest. This price increase was an unprecedented event. The acreage shift estimated to accommodate the increased corn demand was more than 12 million acres (4.9 million hectares) of additional corn with about 7 million acres (2.9 million hectares) coming from lands where soybeans are currently grown (USDA NASS 2007). In addition, a portion of the increased acreage came from existing grasslands or conservation acres. Ultimately, acres planted in corn were predicted to increase by 15 percent between 2006 and 2007, and these additional acres were likely to have a greater environmental impact than the average of existing corn acres. New acres planted to corn may till more marginal soils, and the loss of some soybean rotations, with corn following corn instead of a corn–soybean rotation, requires more fertilizer and more erosive tillage systems to maintain yields (Vyn 2007).

With profits enhanced or ensured by commodity programs, production has also extended into or intensified on marginal lands with related environmental costs. Expansion of wheat production in the West, encouraged by the disaster payments program of the late 1970s, is a classic example. In the more recent past, new soybean price supports along with new technology for soybean production with minimum tillage drove the loss of prairie in the Dakotas.

Budget limitations will undoubtedly be even more important in future farm bill discussions. In the past, under tight budgets, conservation programs suffered more cuts than commodity programs, and this continued into 2006 (Chite 2005). For example, appropriations for the CSP have repeatedly been reduced from authorized levels, in part to pay for disaster relief (Zinn and Cowan 2006). Although the CSP was intended for national implementation, budget limitations and the reallocation of CSP funds to commodity-related programs resulted in implementation of the CSP being delayed by more than two years after the 2002 farm bill passed. By 2006, with a total of three CSP sign-ups in the 282 agricultural watersheds eligible nationally, 15.8 million acres (6.3 million hectares) of working farmland had been enrolled in the CSP (Color Figure 29; Cowan 2006). This is less than half the area enrolled under the Conservation Reserve Program (CRP; about 36 million acres [14.5 million hectares]). Another 1 million acres (0.4 million hectares) are in the Conservation Reserve Enhancement Program (CREP) and the CRP Wetlands Program combined nationally (Zinn and Cowan 2006), with an additional 1.6 million acres (0.6 million hectares) in the Wetlands Reserve Program (WRP; Hanrahan and Zinn 2005). To date, more than twice as much land has been taken out of production and placed in all types of agricultural reserves than the area enrolled in the CSP working lands program. The CSP working lands approach is of particular interest because it is highly consistent with the agricultural practices and patterns included in Scenarios 2 and 3 (Part 2), and the changed agricultural practices described in the MRB integrated assessment (Chapter 15).

Traditionally, the trade-off between production policy and the price of conservation has been important to the overall cost of U.S. agricultural policy (Claassen 2006b). When the market price of commodities has been low and farm incomes have been supported by government payments, higher crop support payments have required larger conservation budgets so that conservation could be bought away from production. If biofuel production results in continuing high commodity prices above the level of government price supports, however, market prices will bid land away from conservation or, at least, reduce the attractiveness of conservation practices on the land. One potential offset of this could occur if commodity prices remain high when the next farm bill is written into law. If this occurs, the savings from reduced commodity payments while prices are high might be directed toward conservation programs, and if conservation program payments are high enough, conservation would be able to bid successfully against high market prices.

WTO-mandated restrictions on commodity programs (Chapter 3) may provoke interest in enhancing conservation programs (Hanrahan and Zinn 2005; USDA NRCS 2006). Under current WTO rules, this would be the case even with the failure of the Doha Round, and it could open the way for considering some of the future scenarios and environmentally beneficial practices described in Part 2 of this book. Currently the WTO clearly allows general service expenditures (such as research and extension) and land retirement by placing land in reserves such as the CRP or the bioreserves in Scenario 3—if land retirement is not tied to a specific commodity or used to compensate farmers specifically for loss of income from complying with environmental programs (USDA NRCS 2006). Decoupled income support, such as the direct payments that were tried under the 1996 farm bill (known as the "Freedom to Farm Act"), are also acceptable. Cross-compliance requirements, like measures to achieve environmental benefits being implemented under the European Union's (EU) Common Agricultural Policy 2000 (CAP) reforms, can be tied to payments as long as they do not violate the decoupled income support rule of the WTO's predecessor, the 1997 General Agreement on Tariffs and Trade (GATT).

Under CAP 2000, European farmers choose from a menu of practices to achieve environmental benefits. Although to date, U.S. policymakers have not embraced "multifunctional" agricultural landscape benefits as defined by the EU, this approach is only one way that U.S. policy could achieve the environmental benefits. As the MRB and Corn Belt futures integrated assessments suggest, modifications to current conservation programs and rules, as well as related federal policies, could also achieve significant environmental benefits. If WTO negotiations seriously affect U.S. agricultural policy, farmers could embrace increased support for environmental benefits as they face mandated declines in commodity supports.

## Adapting Current Policy: Moving toward Visible Change

Current conservation programs and rules, modified in concert with commodity-related programs, could help achieve the environmental results called for by the MRB integrated assessment and demonstrated in Scenarios 2 and 3. Conservation and commodity programs may appear to be independent aspects of agricultural policy, but they actually have competed with each other in farmers' land rent decisions; in congressional appropriations (Womach et al. 2006); and, to some degree, in their effects on environmental quality (Lubowski et al. 2006). If future commodity policies elevate the importance of environmental benefits, several characteristics of current conservation policy (described in the sections that follow) can be instrumental.

### Conservation on Working Lands and Retired Lands

Similar to current federal conservation programs, the Corn Belt scenarios described in Chapter 4 include both working lands programs and land retirement programs. Under current policies, commodity programs affect both types of conservation programs—working lands programs need to incorporate production activities as influenced by commodity programs, and land retirement programs like the CRP need to provide farmers benefits beyond what they would receive if their land were in production with commodity programs. The CRP has amply demonstrated that land retirement can en-

hance soil conservation, water quality, and habitat (Chapter 3). Among the future scenarios, Scenario 3 includes a more predictable and enduring land retirement program
in the form of bioreserves that are permanently allocated to ecosystem restoration as
habitat by public land purchase.

Keeping land in retirement or adding additional acres to such programs can be
difficult when commodity prices or price supports are high. In 2007, a large number
of acres are scheduled to come out of the CRP, making it a critical year for CRP lands.
To meet administrative needs, in 2005 the U.S. Department of Agriculture (USDA) began re-enrolling land from this cohort that had high environmental value. As a result,
80 percent of these acres were re-enrolled when corn prices were at $2/bushel. This
success in re-enrollment would not have been possible after the market price of corn
exceeded $3/bushel in the fall of 2006, reaching $4/bushel by early in 2007. An important question is whether, in the future, the federal government allows enrolled acres to
come out of the various land retirement programs in order to reduce higher commodity prices, which are of concern to livestock producers and the biofuels industry.

### Compliance Monitoring

Compared with land retirement, working lands programs, which include environmentally beneficial landscape patterns and management regimes within fields of commodity production, have less proven broad environmental benefits. As a result, some of
these lands may require more intensive monitoring to ensure compliance. Reduced,
selective use of nitrogen fertilizer inputs (Chapter 15) is a powerful example of working lands conservation, which the alternative Corn Belt futures achieved by somewhat
different combinations of precision farming and reduced area in row crops. Reducing
fertilizer applications on existing field patterns could cumulatively lead to significant
reductions of nitrogen loading to the Gulf of Mexico (Chapter 15), but this would
be considerably more difficult to monitor than the visible landscape patterns of land
retirement or working lands programs that employ small patches of perennial cover,
wetlands, or riparian buffers in a row crop matrix. Some working land programs that
are straightforward to visually monitor, such as the adoption of contour farming or
reduced tillage methods, may deliver environmental benefits such as sediment reduction and carbon sequestration, at lower unit cost than land retirement (Feng et al.
2005). The perennial strip intercropping pervasively employed in Scenario 3 is an example of a directly visible working lands practice that could have multiple benefits by
improving water quality, habitat, and carbon sequestration. In the end, because land
retirement and working lands programs may be more effective in somewhat different
situations, combining both types of programs may be most cost-effective for achieving multiple benefits (Feng and Kling 2005). Monitoring will be critically important
in accomplishing these goals. In addition, monitoring will play a key role in adequate
adaptive management to ensure program success (Cox 2006).

### Cross-Compliance

In the past, cross-compliance set a basic conservation standard that producers were
required to meet to receive production subsidies (Chapter 3). A modified approach
could again become an important conservation tool for working lands. Although the
presumption of a required environmental standard is stronger in Europe than in the
United States today—given the extra leverage of Europe's higher agricultural subsidies—

U.S. conservation programs could be more effective if there were an easily monitored minimum standard requiring all farmers to meet fundamental societal obligations for environmental stewardship (GAO 2003). Much like under the CSP, operators themselves would first have to meet the basic minimum standard on their farms, and then, under a voluntary system, they might be further rewarded for achieving even greater levels of conservation.

## Rules and Regulations

In the past, the lack of strict policy and program definitions has given USDA professionals and state conservationists latitude to make judgments about what improvements fit local conditions. An effective political opposition ultimately resisted more specific rules after conservation compliance was adopted as part of the 1985 farm law; farmers who were out of compliance were held up as symbols of unjust government interference by various farm and property rights groups. Consequently, the overall environmental effect of cross-compliance as it was applied after 1985 was limited. Yet many actions that could advance the environmental benefits of U.S. agricultural policy will require more rather than less specificity about practices implemented and standards met, along with performance monitoring, especially on working lands (Cox 2006). For this reason, there may be some advantage in working lands programs that produce visible changes in landscape patterns, and are consequently more easily monitored using the remote sensing technology that the USDA has employed for decades as part of the commodity programs. Several features of Scenarios 2 and 3 would produce such visible changes—the continuous cover of filter strips, pasture, and hay in Scenario 2; the perennial strips in Scenario 3; and the wetlands and stream buffers in both scenarios.

## Targeting

Targeting has been favored and employed in a variety of ways to make conservation efforts more effective. Numerous studies have repeatedly demonstrated its cost-effectiveness (see, for example, Claassen 2006b). Traditional U.S. policy, however, has achieved dual goals of conservation *and* cash disbursement to rural areas by subsidizing conservation practices that are broadly distributed across the entire nation's agricultural areas. For example, the CRP was first administered so as to reduce production and help farmers and landlords during the farm financial crisis of the 1980s. Initially, the program was targeted only to highly erodible lands, but after 1990 the Environmental Benefits Index (EBI) was used to target a wider range of environmental benefits over a broader geographic area (Feng et al. 2003).

Both recommendations for the MRB (Chapter 15) and all the Corn Belt scenarios (Chapter 4) employ targeting on working lands. For example, all the Corn Belt scenarios target streams for buffers of various dimensions; Scenarios 2 and 3 target perennial agricultural land covers to more erodible lands; Scenario 2 targets wetlands, drainage outlets, and floodplains for detention of runoff to meet water quality goals; and Scenario 3 targets working lands that connect stream corridors and bioreserves for perennial strip intercropping and organic practices (Color Figures 38–43). Analogous to the expansion of CRP targeting after 1990, Scenario 3 targets some land retirement purchases for bioreserves to soils and landscape locations that previously supported these indigenous ecosystems (Color Figures 5 and 15). And the working lands farming practices that Scenario 3 targets, with a goal of connecting the reserves and stream corridors, further enhance habitat values (Color Figures 42 and 43). These connections create a more widely distributed range of geographic targets under Scenario 3 than under Scenario 2. In both scenarios, the riparian and wetland areas recommended to reduce delivery of nitrogen to the Gulf of Mexico (Chapter 15) would be a target of conservation spending. Different benefits are obtained when programs are targeted on specific priorities. Targeting with higher priority for nutrient reduction would increase the opportunity to achieve that goal.

### Performance Standards

Most agricultural conservation programs have relied on best management practices (BMPs) rather than performance standards. From a straightforward economic stand-point, performance standards are appealingly efficient because they make the objective itself the result to be measured. Economists agree with natural and physical scientists that performance standards would be more effective and efficient than relying only on BMPs to achieve environmental benefits. With BMPs, a given practice is essentially assumed to yield a particular environmental result, and problems of measuring the effects of BMPs across large areas compound this problem. Particularly for diffuse, dynamic environmental benefits such as water quality and habitat, however, performance standards are difficult to measure efficiently and reliably in practice. For example, monitoring runoff from individual farms or small subwatersheds, as suggested in Scenario 2, could pose a considerable challenge in instrumentation or modeling. For practical implementation, selected characteristics of either Scenario 2 or 3 probably would need to be translated into a menu of BMPs, much as in CSP implementation today. The important question is: Can representative monitoring be undertaken in a cost-effective way that is sufficient to provide improved ground-truthing of expected outcomes for BMPs?

### Competitive Bidding

Along with targeting and performance standards, carefully implemented competitive bidding could enhance the cost and environmental effectiveness of conservation programs (Claassen 2006b). Although the competitive bidding that was initially employed in early rounds of the CRP program was not demonstrably cost-effective, improved procedures brought about better outcomes later in the process. In the 2002 farm law, though, Congress specifically required that USDA avoid cost criteria.

The efficacy of competitive bidding depends on the structure of the bidding process and the knowledge held by the players. Competitive bidding may be more challenging when funds must be spent within a short time and over a wide range of local conditions, but the potential for significant cost savings makes competitive bidding an important approach to consider in making large-scale conservation programs more cost-effective. When properly structured, bidding can be a low-cost way to obtain conservation benefits. Such a competitive bidding procedure for enrollment in conservation programs should factor in the environmental benefits of conservation practices, which could be modeled as described in Chapters 5–15, to help to establish a rational basis for evaluating comparable bids.

An example of a successful competitive bidding conservation program that encompasses many of the aspects of Scenarios 2 and 3 has been implemented in the province of Victoria, Australia. This program uses geographic information system (GIS) analysis of soil, habitat, and hydrology to determine suitability for program eligibility—much like the Corn Belt watershed futures were developed using GIS coverage of soil and land cover to determine future possible land covers and management practices. Environmental benefits that require management across property boundaries, such as connectivity for habitat restoration or salinity mitigation, are recognized. Farmers bid on implementation of environmentally beneficial practices in a blind process that prevents gaming and allows the government to achieve priority benefits at a fraction of the cost of an incentive program (Eigenram et al. 2006). This is only one example of a farm program that demonstrates that a competitive bidding program that achieves environmental benefits and supports farmers' incomes is a realistic goal. With further innovation in U.S. agricultural policy, this goal could be reached in the United States as well.

### *Farmer Choices*

Finally, we must recognize that the perceptions, values, and choices of farmers are fundamental to the effectiveness of farm programs. Farms, and the producers who make management decisions about them, are the decisive units in current conservation and commodity programs. Individual producers volunteer for conservation or commodity programs. Other land units, such as the watershed, may be critical to achieving environmental benefits, but the land and the producer who manages it is the practical, legal, and traditional unit for decisionmaking. This reflects a cultural and legal tradition in the United States with respect to property rights and the role of individuals in common concerns such as conservation. Compared with some other countries, the United States has a stronger orientation toward the individual operator and land owner.

The Corn Belt futures were evaluated in in-depth interviews with a small number of farmers who would have been knowledgeable about agriculture in the study watersheds (Chapter 6). These farmers perceived both Scenario 2 and Scenario 3 as being valuable for the future of the people of Iowa, given the assumption that none of the policy scenarios would affect their farm income (Chapter 6). For policymakers, determining which alternatives deliver wide societal and environmental benefits in a way that offers farmers adequate financial and personal rewards is crucial to successful change.

## Options for Future Agricultural Policy

We know that agricultural policy, technology, and markets have driven past changes in the Corn Belt, shaping landscape patterns, practices, and environmental and societal impacts (Chapter 2). To address the problem of hypoxia of the Gulf of Mexico, support the restoration of Gulf coastal ecosystems, and create environmental benefits within the Corn Belt itself, several policy approaches are plausible. The integrated assessment of alternative future scenarios (Chapter 14) contributes greatly to envisioning aims for policy that have been recommended for the entire MRB (Chapter 15). These include changes in farming practices, networked wetland and riparian ecosystems, and off-channel storage of agricultural runoff. In fact, many of the pattern and practice ideas in the alternative futures could be recombined in different ways to achieve alternative policy aims, including different approaches to targeting and compliance. And as additional ideas for farm programs are generated, these can be assimilated into the landscape futures to assess their trade-offs. In addition, the broader environmental benefits accruing from the alternative scenarios could be accounted for in dollar terms. The futures approach allows site-specific examination of trade-offs and multiple benefits that extend beyond water quality to the other goals, such as the economic viability of farming.

Even though the alternative futures demonstrate how a different, local frame of analysis can be used to consider these trade-offs, they do not supply detailed road maps for policy implementation. If there were a road map, it would undoubtedly show several alternative paths—each suggesting different advantages and concerns about everything from farm income to time frames for adoption to public and political acceptability. As they stand, the futures challenge policymakers to develop alternative pathways for reaching desirable future landscapes characteristics. We describe several policy approaches in the list that follows (see Table      for a summary).

- *A direct buy-out of existing subsidies*, with producers receiving a large cash payment up front and foregoing the right to future farm program payments. Although this approach would require a huge initial cash expenditure by the government, it would return agriculture's prices and resource use to free market forces. It could be a jarring change to rural economies, with rapid changes in land rents that could dramatically affect the financial planning and well-being of rural land owners. The societal pain associated with this change would very likely tempt Congress to institute commodity payments again later, repeating the experience that occurred when the 1996 farm law decoupled and gradually reduced subsidies from commodities (Chapter 3). This approach would, however, remove artificial incentives for increased production of commodities, even though it would not directly ameliorate the environmental impacts caused by some agricultural production practices.

- *Green conservation payments*. These jarring changes could be largely avoided if green conservation payments were offered instead of commodity payments as the primary means of government support for agriculture (American Farmland Trust 2006a). Properly structured to be easily monitored and have reliable effects, payments for green practices might be very positive for the environment and consistent with WTO definitions of allowable green box policy, as well as providing stable farm incomes. Payments could support the types of practices employed in Scenarios 2 and 3, and resulting gains in quality of life and environmental health could have ripple effects in related community and economic benefits. But existing WTO rules that prohibit compensating farmers for more than the cost of a conservation practice could limit this approach.

- *A market-based land retirement program* in which farmers agree to set aside increasing amounts of land for increased crop price guarantees. With this approach, low-quality land (often with environmental problems for agriculture) gets bid into short-term reserves and the most productive land (sometimes having fewer problems) remains in production. An obstacle to using this approach is that crop price guarantees are inconsistent with WTO mandates. In addition, short-term set-asides could achieve some water quality goals while they are being employed, but would not ensure reduced nitrogen flows to the Gulf or enhanced biodiversity over the longer term. Market-based set-asides would bid in (or out) different pieces and types of land over time for different crops, depending on market prices. In the process, supply and demand would be adjusted. But short-term reserves may be particularly ineffective in ameliorating hypoxia, because of long time lags between reductions in nitrogen loading and reduced nitrogen flows into surface waters (Chapter 15).

- A government-directed *supply control land retirement program*. Reminiscent of programs implemented under traditional USDA set-asides, these programs would require farmers to set aside a percentage of their cropland for a specific crop to get price guarantees for that crop. Traditionally, the secretary of agriculture made the determination about how much corn, wheat, or other cropland would be set aside before opening the program for enrollment. One criticism of this approach is that the secretary's decision may not reflect good knowledge of future weather or market decisions. On the other hand, in practice, farmers set aside their worst land, which serves to protect such areas as farmed wetlands to some extent.

- *Income insurance*. Such an approach is best illustrated by the integrated farm revenue program suggested by the American Farmland Trust (AFT). These types of programs would create an income insurance that would protect farmers against drops in revenue rather than price as existing programs do. Farmers could continue to carry separate crop insurance for their own individual weather risk, but the government-supported income insurance would protect farmers against marketwide risks through a national revenue deficiency program. Such a program is seen as a less expensive way to offer basic income support to farmers. In addition, it would allow more expenditure on a possible companion program of green payments (American Farmland Trust 2006b). An essential point was highlighted by the AFT proposals—income is critical to conservation. Big Hugh Bennett, the first leader of the Soil Conservation Service, recognized that farmers in financial difficulty would mine the natural resources to survive. Adequate income was as critical to conservation in the 1930s as it is today.

Environmental shortcomings of the other approaches suggest serious consideration of a green payments program or an income deficiency program; in either approach funds currently directed toward commodity program payments would be redirected to encouraging conservation. This would be particularly appropriate when commodity prices are high and potential federal commodity payments would be minimal. With increased support for green payments, a wide array of environmental benefits could be achieved on both retired lands and working lands, likely at overall expenditures less than those of current commodity programs. Land needing conservation activities could drawer higher payments than received under current commodity programs, and some land that now receives commodity payments might receive lower conservation payments, depending on how the new programs were targeted. This new distribution could be politically controversial, despite the environmental and economic efficiencies it could create. Incorporating green payments for a wide range of environmentally beneficial activities on working lands, such as those described in Scenarios 2 and 3, could help to more evenly distribute overall farm income. Both the CRP and CSP have been distributed over wider areas when a fuller range of environmental benefits have been considered.

**Table 16-1.** Overview of Potential Federal Policy Approaches to Achieving Environmental Benefits

| Policy option | Advantages | Concerns |
|---|---|---|
| **Direct buyout of existing commodity payments** | Return agriculture to free market conditions | • Large initial cost<br>• Does not directly address environmental impacts<br>• Severe impact on land values<br>• Equity issues for landowners versus land tenants<br>• Likelihood of substitute payments later |
| **Market-based short-term land retirement** | Some environmental benefits, for a limited time | • May be inconsistent with WTO<br>• Short-term set-asides may result in only limited environmental benefits |
| **Supply control short-term land retirement** | Some environmental benefits, for a limited time | • May be inconsistent with WTO<br>• Short-term set-asides may result in only limited environmental benefits |
| **Substitution of green conservation payments for commodity payments** | Return agriculture to free market conditions<br>Environmental benefits at greater economic efficiency<br>Consistent with WTO | • Geographic redistribution of payments could be controversial<br>• Monitoring compliance with some practices is difficult |
| **Income deficiency insurance** | Potentially less costly<br>Frees funds for conservation<br>Less distorting because it is market-priced | • Less generous than existing countercyclical support for some commodities |

## Designing Conservation Programs, Today and Tomorrow

Whether or not future farm laws are more favorable to conservation and the environment, the integrated assessment of the MRB and the alternative futures for local Corn Belt watersheds offer lessons for the design of existing and new conservation programs. Targeting that uses detailed geographic information to consider landscape characteristics and achieve enhanced habitat and water quality benefits and requires all program participants to meet some minimum stewardship standard; competitive bidding for program participation to achieve greater cost-efficiencies; performance standards translated into clearly visible BMPs that are practical to monitor and enforce—all of these could be integrated and acted on as part of the public's investment in agriculture and the environment.

The integrated assessment of the MRB clearly shows that the environmental impacts of Corn Belt agriculture extend into communities, economies, and ecosystems far beyond the region. To a great degree, the solution to hypoxia of the Gulf of Mexico lies in changed Corn Belt farming practices. The scale of federal farm programs, in which the entire American public invests in the health of agriculture and expects that broader societal benefits accrue, is well suited to addressing such continental-scale challenges for the environment and the economy. In addition, America's global trading partners are pushing for federal agricultural policy that would allow support of farmers and environmental benefits but would not allow support of commodity prices. If U.S. agricultural policy responds to these challenges with more substantial innovation, the recommendations for the MRB and for local Corn Belt watersheds hold promise that future agricultural landscapes and communities could be healthier and more desirable places to live. Furthermore, this book suggests many specific practices and policy options that could be combined in new ways to respond to policy needs.

We who have contributed to this book have been encouraged by the willingness of American farmers to innovate, the capability of American research institutions to bridge disciplinary barriers and conceive of new agricultural patterns and practices, and the responsiveness of American businesses to possibilities for technological innovation. Our enormous national propensity to innovate should be taken into account as policymakers consider what agricultural policy could achieve. Innovation will be required to ensure that the environment, rural communities, and agriculture advance together toward their long-term mutual benefit. If policy makes new gains toward joining these dimensions, each essential to societal well-being, both the Corn Belt and the Gulf will be healthier as a result.

# References

Abaidoo, S., and H. Dickinson. 2002. Alternative and Conventional Paradigms: Evidence from Farming in Southwest Saskatchewan. *Rural Sociology* 67(1): 114–130.

Addams, H. 2000. Q Methodology. In *Social Discourse and Environmental Policy: An Application of Q Methodology,* edited by H. Addams and J. Proops. Northampton, MA: Edward Elgar Publishing.

Addiscott, T.M. 1988. Long-term Leakage of Nitrate from Bare Unmanured Soil. *Soil Use Management* 4: 91–95.

Alexander, R.B., R. Smith, and G. Schwartz. 1996. The Regional Transport of Point and Nonpoint-source Nitrogen to the Gulf of Mexico. In *Proceedings of the Gulf of Mexico Hypoxia Management Conference,* Kenner, Louisiana, December 5–6, 1995. Reston, VA: U.S. Geological Survey (USGS).

Allen, J.C., and K. Bernhardt. 1995. Farming Practices and Adherence to an Alternative-Conventional Agricultural Paradigm. *Rural Sociology* 60: 297–309.

American Farm Bureau. 1998. *The Public Image of America's Farmers—Bridging the Perception Gap.* Park Ridge, IL: American Farm Bureau Federation.

American Farmland Trust. 2006a. *Rewarding Farmers and Ranchers for Environmental Stewardship: Green Payments.* Washington, DC: American Farmland Trust. http://www.farmland.org/programs/campaign/documents/AFT_Agenda2007_GreenPaymentsPublicRationale.pdf (accessed December 18, 2006).

———. 2006b. *Integrated Farm Revenue Program.* Washington, DC: American Farmland Trust. http://www.farmland.org/programs/campaign/documents/AFT_RevProtectionbriefing_overview_webNov2006.pdf (accessed December 18, 2006).

Arbuckle, K.E., and J.A. Downing. 2001. The Influence of Watershed Land Use on Lake N:P in a Predominantly Agricultural Landscape. *Limnology and Oceanography* 46: 970–975.

Arcury, T.A. 1990. Environmental Attitude and Environmental Knowledge. *Human Organization* 49(4): 300–304.

Arnold, J.G., J.R. Williams, and D.R. Maidment. 1995. Continuous-time Water and Sediment-routing Model for Large Basins. *Journal of Hydraulic Engineering* 121(2): 171–183.

Babcock, B. 1999. Whither Farm Policy? *Iowa Agricultural Review* 5(4): 1–6.

———. 2001. The Concentration of U.S. Agricultural Subsidies. *Iowa Agricultural Review* 7(4): 8–9.

———. 2002. Local and Global Perspectives on the New U.S. Farm Policy. *Iowa Agricultural Review* 8(3): 1–3.

Baker, J., and J.M. Laflen. 1983. Water Quality Consequences of Conservation Tillage. *Journal of Soil and Water Conservation* 38: 186–193.

Baker, R.H. 1983. *Michigan Mammals.* Detroit, MI: Michigan State University Press.

Bascompte, J., P. Jordano, and J.M. Olesen. 2006. Asymmetric Coevolutionary Networks Facilitate Biodiversity Maintenance. *Science* 312: 431–433.

Becher, K.D., D.J. Schnoebelen, and K.B. Ackers. 2000. Nutrients Discharged to the Mississippi River from Eastern Iowa Watersheds, 1996–1997. *Journal of the American Water Resources Association* 36(1): 161–173.

Beedell, J.D.C., and T. Rehman. 1999. Explaining Farmers' Conservation Behaviour: Why do Farmers Behave the Way They Do? *Journal of Environmental Management* 57:165–176.

———. 2000. Using Social Psychology Models to Understand Farmers' Conservation Behavior. *Journal of Rural Studies* 16: 117–127.

Beeton, A.M. 2002. Large Freshwater Lakes: Present State, Trends, and Future. *Environmental Conservation* 29(1): 21–38.

Belsky, A.J., A. Matzke, and S. Uselman. 1999. Survey of Livestock Influences on Stream and Riparian Ecosystems in the Western United States. *Journal of Soil and Water Conservation* 54: 419–431.

Bergin, T.M., L.B. Best, K.E. Freemark, and K.J. Koehler. 2000. Effects of Landscape Structure on Nest Predation in Roadsides of a Midwestern Agroecosystem: A Multiscale Analysis. *Landscape Ecology* 15: 131–143.

Bergman, K.O., J. Asking, O. Ekberg, H. Ignell, H. Wahlman, and P. Milberg. 2004. Landscape Effects on Butterfly Assemblages in an Agricultural Region. *Ecography* 27: 619–628.

Berrill, M., S. Bertram, and B. Pauli. 1997. Effects of Pesticides on Amphibian Embryos and Tadpoles. In *Amphibians in Decline: Canadian Studies of a Global Problem*, edited by D.M. Green. St. Louis, MO: Society for the Study of Amphibians and Reptiles.

Best, L.B., K.E. Freemark, J.J. Dinsmore, and M. Camp. 1995. A Review and Synthesis of Habitat Use by Breeding Birds in Agricultural Landscapes of Iowa. *American Midland Naturalist* 134: 1–29.

Bierman, V.J., S.C. Hinz, D. Zhu, W.J. Wiseman Jr., N.N. Rabalais, and R.E. Turner. 1994. A Preliminary Mass Balance Model of Primary Productivity and Dissolved Oxygen in the Mississippi River Plume/Inner Gulf Shelf Region. *Estuaries* 17:886–899.

Bishop, R.A. 1981. Iowa's Wetlands. *Proceedings of the Iowa Academy of Science* 88:11–16.

Blair, W.F. 1957. Changes in Vertebrate Populations under Conditions of Drought. *Cold Spring Harbor Symposia on Quantitative Biology* 22:273–275.

Blaustein, A.R., and D.B. Wake. 1990. Declining Amphibian Populations: A Global Phenomenon? *Trends in Ecology and Evolution* 5: 203–204.

Blevins, R.L., G.W. Thomas, and P.L. Cornelius. 1977. Influence of No-tillage and Nitrogen Fertilization on Certain Soil Properties after Five Years of Continuous Corn. *Agronomy Journal* 69: 383–386.

Blevins, R.L., M.S. Smith, W.W. Frye, and P.L. Cornelius. 1983. Changes in Soil Properties after 10 Years in Non-Tilled and Conventionally Tilled Corn. *Soil Tillage Resource* 3: 135–146.

Bonin, J., J.L. DesGranges, J. Rodrigue, and M. Ouellet. 1997. Anuran Species Richness in Agricultural Landscapes of Quebec: Foreseeing Long-Term Results of Road Call Survey. In *Amphibians in Decline: Canadian Studies of a Global Problem*, edited by D.M. Green. St. Louis, MO: Society for the Study of Amphibians and Reptiles.

Boody, G., B. Vondracek, D. Andow, M. Krinke, J. Westra, J. Zimmerman, and P. Welle. 2005. Multifunctional Agriculture in the United States. *Bioscience* 55(1): 27–38.

Bowles, J.B. 1975. *Distribution and Biogeography of Mammals of Iowa*. Lubbock: Texas Tech Press.

Boyer, E.W., C.L. Goodale, N.A. Jaworski, and R.W. Howarth. 2002. Anthropogenic Nitrogen Sources and Relationships to Riverine Nitrogen Export in the Northeastern USA. *Biogeochemistry* 57: 137–169.

Brezonik, P., K. Easter, L. Hatch, D. Mulla, and J. Perry. 1999. Management of Diffuse Pollution in Agricultural Watersheds: Lessons from the Minnesota River Basin. *Water Science and Technology* 39(12): 323–330.

Bultena, L.G., and E.O. Hoiberg. 1983. Factors Affecting Farmers' Adoption of Conservation Tillage. *Journal of Soil and Water Conservation* 38: 281–284.

Burkart, M., D. James, M. Liebman, and C. Herndl. 2005. Impacts of Integrated Crop-Livestock Systems on Nitrogen Dynamics and Soil Erosion in Western Iowa Watersheds. *Journal of Geophysical Research* 110(G1): G01009–G01009.

Burt, W.H. 1972. *Mammals of the Great Lakes Region*. Ann Arbor: University of Michigan Press.

Bystrom, O. 1998. The Nitrogen Abatement Cost in Wetlands. *Ecological Economics* 24: 321–331.

Cain, Z., and S. Lovejoy. 2004. History and Outlook for Farm Bill Conservation Programs. *Choices Magazine*, 4th Quarter, 37–42. http://www.choicesmagazine.org/2004-4/policy/2004-4-09.htm (accessed December 16, 2006).

Caire, W., J.D. Tyler, B.P. Glass, and M.A. Mares. 1989. *Mammals of Oklahoma*. Norman, OK: University of Oklahoma Press.

Cambardella, C.A., T.B. Moorman, D.B. Jaynes, J.L. Hatfield, T.B. Parkin, W.W. Simpkins, and D.L. Karlen. 1999. Water Quality in Walnut Creek Watershed: Nitrate-Nitrogen in Soils, Subsurface Drainage Water, and Shallow Groundwater. *Journal of Environmental Quality* 28: 25–34.

Campbell, T. 2000. *I_EPIC: Users' Guide*. Ames, IA: Iowa State University Center for Agricultural and Rural Development.

Carolan, M.S., D. Mayerfeld, M. Bell, and R. Exner. 2004. Rented Lands: Barriers to Sustainable Agriculture. *Journal of Soil and Water Conservation* 59(4): 70A–75A.

CBO (Congressional Budget Office). 2005. Policies That Distort World Agricultural Trade: Prevalence and Magnitude. Washington, DC: CBO.

CENR (Committee on Environment and Natural Resources Research). 2000. *Integrated Assessment of Hypoxia in the Northern Gulf of Mexico*. Washington, DC: National Science and Technology Council.

Chite, R. 2005. *Agriculture and FY 2005 Budget Reconciliation*. Report number CRS222086. December 23. Washington, DC: Library of Congress Congressional Research Service (CRS).

Chow, T.L., H.W. Rees, and J.L. Daigle. 1999. Effectiveness of Terraces/Grassed Waterway Systems for Soil and Water Conservation: A Field Evaluation. *Journal of Soil and Water Conservation* 54: 577–583.

Chung, S.W., P.W. Gassman, L.A. Kramer, J.R. Williams, and R. Gu. 1999. Validation of EPIC for Two Watersheds in Southwest Iowa. *Journal of Environmental Quality* 28: 971–979.

Clark, W.R., and R.E. Young. 1986. Crop Damage by Small Mammals in No-till Cornfields. *Journal of Soil and Water Conservation* 41: 338–341.

Clausen, H.D., H.B. Holbeck, and J. Reddersen. 2001. Factors Influencing Abundance of Butterflies and Burnet Moths in the Uncultivated Habitats of an Organic Farm in Denmark. *Biological Conservation* 98: 167–178.

Claassen, R. 2000. Compliance Provisions for Soil and Water Conservation. In *Agricultural Resources and Environmental Indicators*. 2000 edition. Washington, DC: U.S. Department of Agriculture (USDA) Economic Research Service (ERS).

———. 2006a. Compliance Provisions for Soil and Water Conservation. In *Agricultural Resources and Environmental Indicators*. 2006 edition. Economic Information Bulletin number EIB-16. Washington, DC: USDA ERS.

———. 2006b. Emphasis Shifts in U.S. Conservation Policy. *Amber Waves*, July 5–10. http://www.ers.usda.gov/AmberWaves/July06SpecialIssue/Features/Emphasis.htm (accessed December 16, 2006).

Claassen, R., and M. Ribaudo. 2006. Conservation Policy Overview. In *Agricultural Resources and Environmental Indicators*. Washington, DC: USDA ERS.

Claassen, R., V. Breneman, S. Bucholtz, A. Cattaneo, R. Johansson, and M. Morehart. 2004. *Environmental Compliance in U.S. Agricultural Policy: Past Performance and Future Potential*. Washington, DC: USDA ERS.

Claassen, R., et al. 2001. *Agri-environmental Policy at the Crossroads: Guideposts on a Changing Landscape*. Washington, DC: USDA ERS.

Clark, W.C., and G. Majone. 1985. The Critical Appraisal of Scientific Enquiries with Policy Implications. *Science, Technology and Human Values* 10(3): 6–19.

Cochrane, W.W., and C.F. Runge. 1992. *Reforming Farm Policy*. Ames, IA: Iowa State University Press.

Coiner, C., J. Wu, and S. Polasky. 2001. Economic and Environmental Implications of Alternative Landscape Designs in the Walnut Creek Watershed of Iowa. *Ecological Economics* 38(1): 119–139.

Collinge, S.K. 1996. Ecological Consequences of Habitat Fragmentation: Implications for Landscape Architecture and Planning. *Landscape and Urban Planning* 36: 59–77.

Collinge, S.K., K.L. Prudic, and J.C. Oliver. 2003. Effects of Local Habitat Characteristics and Landscape Context on Grassland Butterfly Diversity. *Conservation Biology* 17(1): 178–187.

Connell, J.H. 1978. Diversity in Tropical Forests and Coral Reefs. *Science* 199:1302–1309.

Corry, R.C. 2005. The Performance of Landscape Pattern Indices Applied to Fine-Resolution Data Representing a Highly Fragmented Landscape. *Landscape Ecology* 20: 591–608.

Corry, R.C., and J.I. Nassauer. 2002. Managing for Small Patch Patterns in Human-dominated Landscapes: Cultural Factors and Corn Belt Agriculture. In *Integrating Landscape Ecology into Natural Resource Management*, edited by J. Liu and W. Taylor. Cambridge, MA: Cambridge University Press.

———. 2005. Limitations of Using Landscape Pattern Indices to Evaluate the Ecological Consequences of Alternative Plans and Designs. *Landscape and Urban Planning* 72: 265–280.

Cowan, T. 2006. *Conservation Security Program: Implementation and Current Issues*. Washington, DC: Library of Congress CRS.

Cox, C. 2006. *Final Report from the Blue Ribbon Panel Conducting an External Review of the U.S. Department of Agriculture Conservation Effects Assessment Project*. Ankeny, IA: Soil and Water Conservation Society.

Cruse, R.M. 1990. Strip Intercropping. In *Farming Systems for Iowa: Seeking Alternatives*, edited by D. Keeney. Ames, IA: Leopold Center for Sustainable Agriculture.

Dahl, T.E. 1990. Wetlands Losses in the United States, 1780s to 1980s. U.S. Department of the Interior, Fish and Wildlife Service, Washington, DC. Jamestown, ND: Northern Prairie Wildlife Research Center Home Page. http://www.npwrc.usgs.gov/resource/wetlands/wetloss (accessed December 19, 2006).

Dale, V.H., and R.A. Haeuber (eds.). 2001. *Applying Ecological Principles to Land Management*. New York: Springer-Verlag.

Dale, V.H., et al. 2000. Ecological Principles and Guidelines for Managing the Use of Land. *Ecological Applications* 10: 639–670.

Danielson, B.J. 1992. Habitat Selection, Interspecific Interactions and Landscape Composition. *Evolutionary Ecology* 6: 399–411.

Danielson, B.J., and H.R. Pulliam. 1991. Sources, Sinks, and Habitat Selection: A Landscape Perspective on Population Dynamics. *American Naturalist* 137: s50–s66.

Day, J.W. et al. 2005a. Implications of Global Climatic Change and Energy Cost and Availability for the Restoration of the Mississippi Delta. *Ecological Engineering* 24: 253–265.

Day, J.W., W.J. Mitsch, R.R. Lane, and L. Zhang. 2005b. *Restoration of Wetlands and Water Quality in the Mississippi-Ohio-Missouri (MOM) River Basin and Louisiana Delta*. Final report to Louisiana Department of Natural Resources, Baton Rouge, LA.

Di Luzio, M., R. Srinivasan, and J.G. Arnold. 2000. AVSWAT: An ArcView Extension as Tool for the Watershed Control of Point and Non point Sources. In *Proceedings of the 20th ESRI Annual User Conference*, San Diego, CA, June 26–30, 2000. San Diego, CA: ESRI.

Diana, S.G., and V.R. Beasley. 1998. Amphibian Toxicology. In *Status and Conservation of Midwestern Amphibians*, edited by M.J. Lannoo. Iowa City, IA: University of Iowa Press.

Doak, D.F., and L.S. Mills. 1994. A Useful Role for Theory in Conservation. *Ecology* 75: 615–626.

Dodd, C.K. 1993. The Cost of Living in an Unpredictable Environment: The Ecology of Striped Newts *Notophthalmus perstriatus* during a Prolonged Drought. *Copeia*: 605–614.

Doering, O.C. 2002. Economic Linkages Driving the Potential Response to Nitrogen Over-enrichment. *Estuaries* 25(4b): 809–818.

Doering, O., et al. 1999. Evaluation of the Economic Costs and Benefits of Methods for Reducing Nutrient Loads to the Gulf of Mexico. In *Integrated Assessment on Hypoxia in the Gulf of Mexico*. National Oceanic and Atmospheric Administration (NOAA) Coastal Ocean Program Decision Analysis Series No. 20. Silver Spring, MD: NOAA. 115.

Dramstad, W., and C. Sogge. 2003. Developing Indicators for Policy Analysis. In *Agricultural Impacts on Landscapes: Proceedings from NIJOS/OECD Expert Meeting on Agricultural Landscape Indicators*, Oslo, Norway, October 7–9, 2002. NIJOS Report number 07/2003. Oslo: Norwegian Institute for Land Inventory NIJOS.

Duffy, M., and D. Smith, 2004. *Farmland Ownership and Tenure in Iowa 1982–2002: A Twenty-Year Perspective*. Ames, IA: Iowa State University Extension, PM 1983.

Dunning, J.B., B.J. Danielson, and H.R. Pulliam. 1992. Ecological Processes that Affect Populations in Complex Landscapes. *Oikos* 65: 169–175.

Dutcher, D.D., J.C. Finley, A.E. Luloff, and J. Johnson. 2004. Landowner Perceptions of Protecting and Establishing Riparian Forests: A Qualitative Analysis. *Society and Natural Resources* 17: 329–342.

Duvick, D.N., and T.J. Blasing. 1981. A Dendroclimatic Reconstruction of Annual Precipitation Amounts in Iowa since 1680. *Water Resources Research* 17: 1183–1189.

Egan, K., J. Herriges, C. Kling, J. Downing. 2005. Recreation Demand Using Physical Measures of Water Quality. Department of Economics working paper, Center for Agricultural and Rural Development, Iowa State University. http://www.card.iastate.edu/publications/synopsis.aspx?id=555 (accessed December 18, 2006).

Eghball, B., J.E. Gilley, L.A. Kramer, and T.B. Moorman. 2000. Narrow Grass Hedge Effects on Phosphorus and Nitrogen in Runoff Following Manure and Fertilizer Application. *Journal of Soil and Water Conservation* 55: 172–176.

Ehley, A.M. 1992. Integrated Roadside Vegetation Management: A County Approach to Roadside Management in Iowa. In *Proceedings of the 12th North American Prairie Conference*, 159–160. Cedar Falls, IA: University of Northern Iowa.

Eigenram, M., L. Strappazzon, N. Lansdall, A. Ha, C. Beverly, and J. Todd. 2006. Eco Tender: Auction for Multiple Environmental Outcomes. In *National Action Plan for Salinity and Water Quality*. National Market Based Instruments Pilot Program, Project Final Report, Department of Primary Industries, Government of Victoria, Melbourne.

Eilers, L., and D. M. Roosa. 1994. *The Vascular Plants of Iowa*. Iowa City, IA: University of Iowa Press.

Erlich, P.R. 1988. The Loss of Diversity: Causes and Consequences. In *Biodiversity*, edited by E.O. Wilson. Washington, DC: National Academy of Sciences.

Ervin, A. C., and D. E. Ervin. 1982. Factors Affecting the Use of Soil Conservation Practices: Hypotheses, Evidence and Policy Implications. *Land Economics* 58: 271–292.

European Communities. 2001. *Environment 2010: Our Future, Our Choice*. Luxembourg: European Communities. http://ec.europa.eu/environment/newprg/pdf/6eapbooklet_en.pdf (accessed November 21, 2006).

Exner, D. N., D.G. Davidson, M. Ghaffarzadeh, and R.M. Cruse. 1999. Yields and Returns from Strip Intercropping on Six Iowa Farms. *American Journal of Alternative Agriculture* 14: 69–77.

Fairweather, J.R., and S.R. Swaffield. 2000. Q Method Using Photographs to Study Perceptions of the Environment in New Zealand. In *Social Discourse and Environmental Policy: An Application of Q Methodology*, edited by H. Addams and J. Proops. Northhampton, MA: Edward Elgar Publishing.

Farrar, D. 1981. Perspectives on Iowa's Declining Flora and Fauna—A Symposium. *Proceedings of the Iowa Academy of Science* 88: 1–6.

Feather, P., D. Hellerstein, and L. Hansen. 1999. *Economic Valuation of Environmental Benefits and the Targeting of Conservation Programs: The Case of the CRP*. Agricultural Economics Report No. 778, April. Washington, DC: USDA ERS.

Featherstone, M.A., and B.K. Goodwin. 1993. Factors Influencing a Farmer's Decision to Invest in Long-Term Conservation Improvements. *Land Economics* 69: 67–81.

FDIC (Federal Deposit Insurance Corporation). 2004. Iowa State Profile. Winter.

Feng, H. 2002. *Green Payments and Dual Policy Goals*. Ames, IA: Iowa State University Center for Agricultural and Rural Development (CARD).

Feng, H., and C. L. Kling. 2005. *The Consequences of Co-benefits for the Efficient Design of Carbon Sequestration Programs*. Ames, IA: Iowa State University CARD.

Feng, H., C.L. Kling, L.A. Kurkalova, and S. Secchi. 2003. *Subsidies! The Other Incentive-Based Instrument: The Case of the Conservation Reserve Program*. Ames, IA: Iowa State University CARD.

Feng, H., L.A. Kurkalova, C.L. Kling, and P.W. Gassman. 2005. *Economic and Environmental Co-benefits of Carbon Sequestration in Agricultural Soils: Retiring Agricultural Land in the Upper Mississippi River Basin*. Ames, IA: Iowa State University CARD.

Fletcher, S.R. 2005. *Global Climate Change: The Kyoto Protocol*. Washington, DC: Library of Congress CRS.

Flora, C.B. 2001. *Interactions between Agroecosystems and Rural Communities*. Boca Raton, FL: CRC Press.

Foster J., and M.S. Gaines. 1991. The Effects of a Successional Habitat Mosaic on a Small Mammal Community. *Ecology* 72: 1358–1373.

Freckmann, R.W. 1966. The Prairie Remnants of the Ames Area. *Proceedings of the Iowa Academy of Science* 73:126–136.

Freemark, K. 1995. Assessing Effects of Agriculture on Terrestrial Wildlife: Developing a Hierarchical Approach for the U.S. EPA. *Landscape and Urban Planning* 31: 99–115.

Gagnon, S., J. Makuch, and T.J. Sherman. 2004. *Environmental Effects of US Department of Agriculture Conservation Programs: A Conservation Effects Assessment Bibliography.* Beltsville, MD: Water Quality Information Center, National Agricultural Library, USDA Agricultural Research Service (ARS).

Galatowitsch, S.M., and A.G. van der Valk. 1994. *Restoring Prairie Wetlands: An Ecological Approach.* Ames, IA: Iowa State University Press.

———. 1996. The Vegetation of Restored and Natural Prairie Wetlands. *Ecological Applications* 6(1): 102–112.

GAO (General Accounting Office). 2003. *Agricultural Conservation: USDA Needs to Better Ensure Protection of Highly Erodible Cropland and Wetlands.* Washington, DC: GAO.

Gibbs, J.P. 1993. Importance of Small Wetlands for the Persistence of Local Populations of Wetland-Associated Animals. *Wetlands* 13: 25–31.

Gilliam, J.W., and G.D. Hoyt. 1987. Effect of Conservation Tillage on Fate and Transport of Nitrogen. In *Effects of Conservation Tillage on Groundwater Quality: Nitrates and Pesticides,* edited by T.J. Logan, J.M. Davidson, J.L. Baker, and M.R. Overcash. Chelsea, MI: Lewis Publishers, Inc.

Gollehon, N., M. Caswell, M. Ribaudo, R. Kellogg, C. Lander, and D. Letson. 2001. *Confined Animal Production and Manure Nutrients.* Agriculture Information Bulletin, number AIB771. Washington, DC: USDA ERS. http://www.ers.usda.gov/publications/aib771/ (accessed December 18, 2006).

Goodwin, B.K., and V.H. Smith. 2003. An Ex Post Evaluation of the Conservation Reserve, Federal Crop Insurance, and Other Government Programs: Program Participation and Soil Erosion. *Journal of Agricultural and Resource Economics* 28(2): 201–216.

Goolsby, D.A., W.A. Battaglin, G.B. Lawrence, R.S. Artz, B.T. Aulenbach, and R.P. Hooper. 1999. *Flux and Sources of Nutrients in the Mississippi-Atchafalaya Basin: Topic 3 Report for the Integrated Assessment on Hypoxia in the Gulf of Mexico.* Silver Spring, MD: NOAA Coastal Ocean Office. Decision Analysis Series number 17. http://www.nos.noaa.gov/Products/pubs_hypox.html#Topic3 (accessed December 12, 2006).

Goolsby, D.A. and W.A. Battaglin. 2001. Long-Term Changes in Concentrations and Flux of Nitrogen in the Mississippi River Basin, USA. *Hydrologic Processes* 15: 1209–1226.

Goolsby, D.A., W.A. Battaglin, B.T. Aulenbach, and R.P. Hooper. 2001. Nitrogen Input to the Gulf of Mexico. *Journal of Environmental Quality* 30: 329–336.

Goudy, W., and S.C. Burke. 1994. *Iowa's Counties: Selected Population Trends, Vital Statistics, and Socioeconomic Data.* Ames, IA: Census Services, Iowa State University.

Gould, W.B., W.E. Saupe, and R.M. Klemme. 1989. Conservation Tillage: The Role of Farm and Operator Characteristics and Perception of Soil Erosion. *Land Economics* 65: 167–181.

Gowda, P., A. Ward, D. White, J. Lyon, and E. Desmond. 1999. The Sensitivity of ADAPT Model Predictions of Streamflows to Parameters Used to Define Hydrologic Response Units. *Transactions of the American Society of Agricultural Engineers* 42: 381–389.

Gray, L.C. 1933. *History of Agriculture in the Southern United States to 1860.* Vols. 1 and 2. Washington, DC: The Carnegie Institute of Washington Publishers, 430.

Greeley, W.B. 1925. The Relation of Geography to Timber Supply. *Economic Geography* 1: 1–14.

Gren, I.M. 1993. Alternative Nitrogen Reduction Policies in the Mälar Region, Sweden. *Ecological Economics* 7: 159–172.

Gren, I.M., K. Elofsson, and P. Jannke. 1997. Cost-effective Nutrient Reductions to the Baltic Sea. *Environmental Research Economics* 10: 341–362.

Guru, M., and J. E. Horne. 2000. *The Ogallala Aquifer.* The Kerr Center for Sustainable Agriculture. http://www.kerrcenter.com/ (accessed December 18, 2006).

Haddad, N.M., and K.A. Baum. 1999. An Experimental Test of Corridor Effects on Butterfly Densities. *Ecological Applications* 9(2): 623–633.

Hahne, H.C.H., W. Kroontije, and J.A. Lutz Jr. 1977. Nitrogen Fertilization. I. Nitrate Accumulation and Losses under Continuous Corn Cropping. *Journal of the Soil Science Society of America* 41: 562–567.

Hallberg, G.R. 1989. Nitrate in Ground Water in the United States. In *Nitrogen Management and Ground-Water Protection: Developments in Agricultural and Managed-Forest Ecology 21*, edited by R.F. Follet. New York: Elsevier.

Halley, J.M., R.S. Oldham, and J. W. Arntzen. 1996. Predicting the Persistence of Amphibian Populations with the Help of a Spatial Model. *Journal of Applied Ecology* 33: 455–470.

Hamblin, A. 2000. *Visions of Future Landscapes*. Proceedings of 1999 Australian Academy of Science Fenner Conference on the Environment, May 2–5, 1999, Canberra. Bureau of Rural Sciences, Canberra.

Hanrahan, C.E., and J. Zinn. 2005. *Green Payments in U.S. and European Union Agricultural Policy*. CRS Report RL32624. Washington, DC: Library of Congress CRS.

Hansen, L. 2006. *Wetlands Status and Trends*. Washington, DC: USDA ERS: 42–29, Chapter 2.3.

Hart, C.E., and B. Babcock. 2002. *U.S. Farm Policy and the World Trade Organization: How Do They Match Up?* Ames, IA: Iowa State University Center for Agricultural and Rural Development, 21.

Hatfield, J. L., D.B. Jaynes, M.R. Burkhart, C.A. Cambardella, T.B. Moorman, J.H. Prueger, and M.A. Smith. 1999. Water Quality in Walnut Creek Watershed: Setting and Farming Practices. *Journal of Environmental Quality* 28: 11–24.

Hazard, E.B. 1982. *The Mammals of Minnesota*. Minneapolis: University of Minnesota Press.

Hecnar, S.J. 1997. Amphibian Pond Communities in Southwestern Ontario. In *Amphibians in Decline: Canadian Studies of a Global Problem*, edited by D.M. Green. St. Louis, MO: Society for the Study of Amphibians and Reptiles.

Heimlich, R. 2000. *Agricultural Resources and Environmental Indicators: Overview of Conservation Programs and Expenditures*. Washington, DC: USDA ERS.

———. 2003. *Agricultural Resources and Environmental Indicators*. Washington, DC: USDA ERS.

Heimlich, R., D. Gadsby, R. Claassen, and K. Wiebe. 2000. *Agricultural Resources and Environmental Indicators: Wetlands Programs*. Washington, DC: USDA ERS.

The Heinz Center (The H. John Heinz III Center for Science, Economics and the Environment). 2002. *The State of the Nation's Ecosystems*. Cambridge, UK: Cambridge University Press.

Helgen, J., R.G. McKinnell, and M.C. Gernes. 1998. Investigation of Malformed Northern Leopard Frogs in Minnesota. In *Status and Conservation of Midwestern Amphibians*, edited by M.J. Lannoo. Iowa City, IA: University of Iowa Press.

Hellerstein, D. 2006. *USDA Land Retirement Programs*. Washington, DC: USDA ERS: 175–183, Chapter 5.2.

Hellerstein, D., et al. 2002. *Farmland Protection: The Role of Public Preferences for Rural Amenities*. Washington, DC: USDA ERS.

Henry, A.C., D.A. Hosack, C.W. Johnson, D. Rol, and G. Bentrup. 1999. Conservation Corridors in the United States: Benefits and Planning Guidelines. *Journal of Soil and Water Conservation* 54: 645–650.

Herkert, J.R. 1991. Prairie Birds of Illinois: Population Response to Two Centuries of Habitat Change. *Illinois Natural History Survey Bulletin* 34: 393–399.

Hill, B. 1999. Farm Household Income: Perceptions and Statistics. *Journal of Rural Studies* 15(3): 345–358.

Hinkle, M. K. 2000. Time Out for Genetically Modified Crops. *Journal of Soil and Water Conservation* 55: 111.

Hoppe, R.A., and D.E. Banker. 2006. *Structure and Finances of U.S. Farms: 2005 Family Farm Report*. Washington, DC: USDA ERS. http://www.ers.usda.gov/publications/EIB12/ (accessed December 19, 2006).

House, R., M. Peters, and H. McDowell. 1999. *USMP Regional Agricultural Model*. Unpublished Technical Document. Washington, DC: USDA ERS.

Howarth, R.W., et al. 1996. Regional Nitrogen Budgets and Riverine N & P Fluxes for the Drainages to the North Atlantic Ocean: Natural and Human Influences. *Biogeochemistry* 35: 75–139.

Hudson, D., D. Hite, and T. Hobb. 2005. Public Perception of Agricultural Pollution and Gulf of Mexico Hypoxia. *Coastal Management* 33: 25–36.

Humphreys, A.A., and H.L. Abbot. 1876. (Originally published in 1861.) *Report Upon the Physics and Hydraulics of the Mississippi River; Upon the Protection of the Alluvial Region Against Overflow; and Upon the Deepening of the Mouths*. Professional Papers No. 4. Corps of Topographical Engineers, United States Army War Department.

Huston, M. 1979. A General Hypothesis of Species Diversity. *The American Naturalist* 113: 81–101.

Iowa Cooperative Soil Survey. 1999. http://icss.agron.iastate.edu (accessed December 14, 2006).

Iowa DNR (Department of Natural Resources). 2005. State Preserves. http://sargasso.gis.iastate.edu/preserves/ (accessed November 4, 2005).

———. 2006. Iowa Department of Natural Resources, Iowa Natural Areas Inventory. http://csbweb.igsb.uiowa.edu/imsgate/maps/natural_areas.asp (accessed December 15, 2006).

Iowa State University. 1996. Iowa Soil Properties and Interpretations Database (ISPAID). Ames, IA: Iowa State University.

———. 2004. Iowa Soil Properties and Interpretations Database. Ames, IA: Iowa State University.

Isenhart, T.M., R.C. Schultz, I.P. Colletti, and C.A. Rodrigue. 1995. Design, Function, and Management of Integrated Riparian Management Systems. In *Proceedings of the National Symposium on Using Ecological Restoration to Meet Clean Water Act Goals*, Chicago, IL., 93–101. U.S. Environmental Protection Agency (U.S. EPA).

ISPAID. 2004. http://extension.agron.iastate.edu/soils/pdfs/ISP71MAN.pdf (accessed December 15, 2006). Iowa State University. Iowa Agriculture and Home Economics Experiment Station University Extension Service.

Jackson, H.H.T. 1961. *Mammals of Wisconsin*. Madison, WI: University of Wisconsin Press.

Jackson, L.S., C.A. Thompson, and J.J. Dinsmore. 1996. *The Iowa Breeding Bird Atlas*. Iowa City, IA: University of Iowa Press.

Jackson, L.L., D.R. Keeney, and E.M. Gilbert. 2000. Swine Manure Management Plans in North-Central Iowa: Nutrient Loading and Policy Implications. *Journal of Soil and Water Conservation* 55: 205–212.

Jackson, D.L., and L.L. Jackson. 2002. *The Farm as Natural Habitat*. Washington, DC: Island Press.

Jacobs, J.J., and J.F. Timmons. 1974. An Economic Analysis of Agricultural Land Use Practices to Control Water Quality. *American Journal of Agricultural Economics* 56: 791–798.

Jacobsen, R.L., N. Albrecht, and K.E. Bolin. 1992. Wildflower Routes: Benefits of a Management Program for Minnesota Right-of-Way Prairies. In *Proceedings of the 12th North American Prairie Conference*, 153–158. Cedar Falls, IA: University of Northern Iowa.

Jaynes, D. B., J.L. Hatfield, and D.W. Meek. 1999. Water Quality in Walnut Creek Watershed: Herbicides and Nitrate in Surface Waters. *Journal of Environmental Quality* 28 (1): 45–59.

Jaynes, D.B., D.L. Dinnes, D.W. Meek, D.L. Karlen, C.A. Cambardella, and T.S. Colvin. 2004. Using the Late Spring Nitrate Test to Reduce Nitrate Loss within a Watershed. *Journal of Environmental Quality* 33: 669–677.

John-Alder, H.B., and P. Morrin. 1990. Effects of Larval Density on Jumping Ability and Stamina in Newly Metamorphosed *Bufo woodhousii fowleri*. *Copiea* 1990(3): 856–860.

Johnson, B.A. 2004. *Implementing the Conservation Security Program*. CRS Report for Congress. http://www.ncseonline.org/NLE/CRSreports/04nov/RS21740.pdf, updated November 2004 (accessed December 19, 2006).

Johnson-Groh, C.L. 1985. Vegetation Communities of Ledges State Park, Boone County, Iowa. *Proceedings of the Iowa Academy of Science* 92 (4): 129–136.

Johansson, R. 2006. *Working-Land Conservation Programs*. USDA ERS: 194–201, Chapter 5.4.

Jones, J.R., B.P. Borofka, and R.W. Bachmann. 1976. Factors Affecting Nutrient Loads in Some Iowa Streams. *Water Research* 10:117–122.

Jones, C.A., J.R. Williams, A.N. Sharpley, and C.V. Cole. 1985. Testing the Nutrient Components of EPIC. In *Symposium Proceedings of the Natural Resources Modeling Symposium*, Pingree Park, CO, October 16–21, 1983, edited by D.G. Decoursey. Ft. Collins, CO: USDA ARS.

Jordan, T.E, D.L. Correll, and D.E. Weller. 1997. Relating Nutrient Discharges from Watersheds to Land Use and Streamflow Variability. *Water Resources Research* 33: 2579–2590.

Justic, D., N.N. Rabalais, and R.E. Turner. 1996. Effects of Climate Change on Hypoxia in Coastal Waters: A Doubled $CO_2$ Scenario for the Northern Gulf of Mexico. *Limnology and Oceanography* 41: 992–1003.

———. 2003a. Simulated Responses of the Gulf of Mexico Hypoxia to Variations in Climate and Anthropogenic Nutrient Loading. *Journal of Maritime Systems* 42: 115–126.

———. 2003b. Climatic Influences on Riverine Nitrate Flux: Implications for Coastal Marine Eutrophication and Hypoxia. *Estuaries* 26: 1–11.

———. 2005. Coupling between Climate Variability and Coastal Eutrophication: Evidence and Outlook for the Northern Gulf of Mexico. *Journal of Sea Research* 54: 25–35.

Kanwar, R.S., J.L. Baker, and J.M. Laflen. 1985. Nitrate Movement through the Soil Profile in Relation to Tillage System and Fertilization Application Method. *Transactions of the American Society of Agricultural Engineers* 28: 1802–07.

Karl, T.R., and W.E. Riebsame. 1984. The Identification of 10- to 20-Year Temperature and Precipitation Fluctuations in the Contiguous United States. *Journal of Climate and Applied Meteorology* 23: 950–966.

Karlen D.L., D.L. Dinnes, D.B. Jaynes, C.R. Hurburgh, C.A. Cambardella, T.S. Colvin, and G.R. Rippke. 2005. Corn Response to Late-spring Nitrogen Management in the Walnut Creek Watershed. *Agronomy Journal* 97 (4): 1054–1061.

Keeney, D.R. 1986. Sources of Nitrate to Ground Water. *Critical Reviews in Environment Control* 16: 257–304.

Kent, T.H., and J. J. Dinsmore. 1996. *Birds in Iowa*. Ames and Iowa City, IA: Published by the authors.

Kitur, B.K., M.S. Smith, R.L. Blevins, and W.W. Frye. 1984. Fate of N-Depleted Ammonium Nitrate Applied to No-tillage and Conventional Tillage Corn. *Agronomy Journal* 76: 240–242.

Kline, J., and D. Wichelns. 1996. Empirical Evidence of Public Preferences for Farmland Preservation. In *Environmental Enhancement through Agriculture*, edited by W. Lockeretz. Ames, IA: Iowa State University Press.

Kling, C., S. Secchi, L. Kurkalova, P. Gassman, H. Feng, and M. Jha. 2005. Natural Resources at Risk: Water Quality and the Dead Zone in the Gulf of Mexico. Paper presented at the AERE (Association of Environmental and Resource Economists) Workshop, Jackson, WY.

Knutson, M.G., J.R. Sauer, D.A. Olsen, M.J. Mossman, L.M. Hemesath, and M.J. Lannoo. 1999. Effects of Landscape Composition and Wetland Fragmentation on Frog and Toad Abundance and Species Richness in Iowa and Wisconsin, U.S.A. *Conservation Biology* 13:1437–1446.

Knutson, M.G., W.B. Richardson, D.M. Reineke, B.R. Gray, J.R. Parmelee, and S.E. Weick. 2004. Agricultural Ponds Support Amphibian Populations. *Ecological Applications* 14:669–684.

Kronvang, B., R. Grant, S.E. Larsen, L.M. Svendsen, and H.E. Andersen. 1995. Non-point-source Nutrient Losses to the Aquatic Environment in Denmark, Impact of Agriculture. *Marine and Freshwater Research* 46:167–177.

Kross, B.C., G.R. Hallberg, D.R. Brunner, K. Cherryholmes, and J.K. Johnson. 1993. The Nitrate Contamination of Private Well Water in Iowa. *American Journal of Public Health* 83: 270–272.

Kucera, C.L. 1952. An Ecological Study of a Hardwood Forest Area in Central Iowa. *Ecological Monographs* 22(4): 283–299.

Lammers, T.G., and A.G. van der Valk. 1977a. A Checklist of the Aquatic and Wetland Vascular Plants of Iowa: 1. Ferns, Fern Allies and Dicotyledons. *Proceedings of the Iowa Academy of Science* 84(2): 41–88.

———. 1977b. A Checklist of the Aquatic and Wetland Vascular Plants of Iowa: 2. Monocotyledons; Plus a Summary of the Geographic and Habitat Distribution of All Aquatic and Wetland Species in Iowa. *Proceedings of the Iowa Academy of Science* 85 (4): 121–163.

Lane, R., J.W. Day, and B. Thibodeaux. 1999. Water Quality Analysis of a Freshwater Diversion at Caernarvon, Louisiana. *Estuaries* 22: 327–336.

Leue, A. 1886. *The Forestal Relations of Ohio*. First Annual Report of the Ohio State Forestry Bureau. Columbus, OH: Westbote Co., State Printers.

Lewandrowski, J., M. Peters, C. Jones, R. House, M. Sperow, M. Eve, and K. Paustian. 2004. *Economics of Sequestering Carbon in the U.S. Agricultural Sector*. Washington, DC: USDA ERS.

Lichtenberg, E., and R. Zimmerman. 1999. Information and Farmers' Attitudes about Pesticides, Water Quality, and Related Environmental Effects. *Agriculture, Ecosystems and Environment* 73:227–236.

Liebman, M. 2001. Managing Weeds with Insects and Pathogens. In Ecological Management of Agricultural Weeds, edited by M. Liebman, C.L. Mohler, and C.P. Staver. Cambridge, UK: Cambridge University Press.

Likens, G.E., F.H. Bormann, R.S. Pierce, J.S. Eaton, and N.M. Johnson. 1977. *Biogeochemistry of a Forested Ecosystem*. New York: Springer-Verlag.

Lockeretz, W. (ed.). 1997. *Visions of American Agriculture*. Ames, IA: Iowa State University Press.

Löfgren, S., A. Gustafson, S. Steineck, and P. Ståhlnacke. 1999. Agricultural Development and Nutrient Flows in the Baltic States and Sweden after 1988. *Ambio* 28: 320–327.

Lowrance, R.R. 1992. Nitrogen Outputs from a Field-size Agricultural Watershed. *Journal of Environmental Quality* 21: 602–607.

Lubowski, R.N., S. Bucholz, R. Claassen, M.J. Roberts, J.C. Cooper, A. Gueroguieva, and R. Johansson. 2006. Environmental Effects of Agricultural Land Use Change: Economics and Policy. Washington, DC: USDA ERS.

Ludwig, J.A. 1999. Disturbance and Landscapes: The Little Things Count. In *Issues in Landscape Ecology*, edited by J.A.Wiens and M.R. Moss. Guelph, Ontario: The International Association for Landscape Ecology.

Lyons, J., B. M. Weigel, L.K. Paine, and D.J. Undersander. 2000. Influence of Intensive Rotational Grazing on Bank Erosion, Fish Habitat Quality, and Fish Communities in Southwestern Wisconsin Trout Streams. *Journal of Soil and Water Conservation* 55: 271–276.

Manale, A. 2000. Flood and Water Quality Management through Targeted, Temporary Restoration of Landscape Functions: Paying Upland Farmers to Control Runoff. *Journal of Soil and Water Conservation* 55: 285–295.

Manguerra, H.B., and B.A. Engel. 1998. Hydrologic Parameterization of Watersheds for Runoff Prediction Using SWAT. *Journal of the American Water Resources Association* 34 (5): 1149–1162.

Mann, C.C., and M.L. Plummer. 1995. *Noah's Choice: The Future of Endangered Species*. New York: Knopf.

Mapp, H.P., D.J. Bernardo, G.J. Sabbagh, S. Geleta, and K.B. Watkins. 1994. Economic and Environmental Impacts of Limiting Nitrogen Use to Protect Water Quality: A Stochastic Regional Analysis. *American Journal of Agricultural Economics* 76: 889–903.

Martin, L.M., K.A. Moloney, and B.J. Wilsey. 2005. An Assessment of Grassland Restoration Success Using Species Diversity Components. *Journal of Applied Ecology* 42: 327–336.

Matson, P.A., W.J. Parton, A.G. Power, and M.J. Swift. 1997. Agricultural Intensification and Ecosystem Properties. *Science* 277: 504–508.

May, R.M. 1975. Biological Populations Obeying Difference Equations: Stable Points, Stable Cycles, and Chaos. *Journal of Theoretical Biology* 51: 511–524.

McCann, E., S. Sullivan, D. Erickson, and R. De Young. 1997. Environmental Awareness, Economic Orientation, and Farming Practices: A Comparison of Organic and Conventional Farmers. *Environmental Management* 21: 747–758.

McConnell, V., and M. Walls. 2005. The Value of Open Space: Evidence from Studies of Nonmarket Benefits. *Resources for the Future*. http://www.rff.org/rff/Documents/RFF-REPORT-Open%20Spaces.pdf (accessed December 19, 2006).

McHenry, H. 1997. Wild Flowers in the Wrong Field are Weeds! Examining Farmers' Constructions of Conservation. *Environment and Planning A* 29:1039–1053.

McIsaac, G.F., M.B. David, G.Z. Gertner, and D.A. Goolsby. 2001. Nitrate Flux in the Mississippi River. *Nature* 414: 166–167.

———. 2002. Relating Net Nitrogen Input in the Mississippi River Basin to Nitrate Flux in the Lower Mississippi River: A Comparison of Approaches. *Journal of Environmental Quality* 31: 1610–1622.

McIsaac, G.F., and X. Hu. 2004. Net N Input and Riverine N Export from Illinois Agricultural Watersheds with and without Extensive Tile Drainage. *Biogeochemistry* 70: 251–271.

McMahon, M.A., and G.W. Thomas. 1976. Anion Leaching in Two Kentucky Soils under Conventional Tillage and Killed Sod Mulch. *Agronomy Journal* 71: 1009–1015.

McNeeley, J.A., M. Gadgil, C. Leveque, and K. Redford. 1995. Human Influences on Biodiversity. In *Global Biodiversity Assessment*, edited by V.A. Heywood. Cambridge, UK: Cambridge University Press.

Menzel, B. W., J. B. Barnum, and L. M. Antosch. 1984. Ecological Alterations of Iowa Prairie-Agricultural Streams. *Iowa State Journal of Research* 59: 5–30.

Menzies, D. 2000. Clean and Green? Environmental Quality on the Dairy Farm. Ph. D. Thesis, Landscape Architecture, Lincoln University, New Zealand.

Merrigan, K.A. 1997. Government Pathways to True Food Security. *In Visions of American Agriculture*, edited by W. Lockeretz. Ames, IA: Iowa State University Press.

Milham, N. 1994. An Analysis of Farmers' Incentives to Conserve or Degrade the Land. *Journal of Environmental Management* 40: 51–64.

Millennium Assessment 2005. Millennium Ecosystem Assessment. http://www.millenniumassessment.org/en/index.aspx (accessed September 14, 2006).

Miller M.F., and H.H. Krusekoff. 1932. The Influence of Systems of Cropping and Methods of Culture on Surface Runoff and Soil Erosion. *Missouri Agricultural Experiment Station Research Bulletin 177*.

Mitchell, R. B., et al. 2006. Information and Influence. In *Global Environmental Assessments: Information and Influence*, edited by R.B. Mitchell, W.C. Clark, D.W. Cash, and N.M. Dickson. Cambridge, MA: MIT Press.

Mitsch, W.J., J.W. Day, J.W. Gilliam, P.M. Groffman, D. L. Hey, G.W. Randall, and N. Wang. 1999. *Reducing Nutrient Loads, Especially Nitrate-Nitrogen, to Surface Water, Groundwater, and the Gulf of Mexico: Topic 5 Report for the Integrated Assessment on Hypoxia in the Gulf of Mexico*. Decision Analysis Series no. 19. Silver Spring, MD: NOAA Coastal Ocean Office.

———. 2001. Reducing Nitrogen Loading to the Gulf of Mexico from the Mississippi River Basin: Strategies to Counter a Persistent Ecological Problem. *BioScience* 51(5): 373–388.

Mohanty, S., and P. Kaus. 1999. European Union Agricultural Reforms: Impacts for Iowa. *Iowa Agricultural Review* 5(4): 8-9.

Moyer, W., and T. Josling. 2002. *Agricultural Policy Reform: Politics and Process in the EU and US in the 1990s*. Burlington, VT: Ashgate Publishing Company.

Mueller, D.K., P.A. Hamilton, D.R. Helsel, K.J. Hitt, and B.C. Ruddy. 1995. *Nutrients in Ground Water of the United States — An Analysis of Data through 1992*. Water-Resources Investigations Report No. 95-4031. U.S. Geological Survey, Denver, CO.

Muir, J. 1965. *The Story of My Boyhood and Youth*. Madison, WI: University of Wisconsin Press.

Munger, R., et al. 1997. Intrauterine Growth Retardation in Iowa Communities with Herbicide Contaminated Drinking Water Supplies. *Environmental Health Perspectives* 105(3): 308–314.

Napier, T.L., and D.E. Brown. 1993. Factors Affecting Attitudes toward Groundwater Pollution among Ohio Farmers. *Journal of Soil and Water Conservation* 48: 432–439.

Napier, T.L., M. Tucker, and S. McCarter. 2000. Adoption of Conservation Production Systems in Three Midwest Watersheds. *Journal of Soil and Water Conservation* 55 (2): 123–134.

Napier, T.L., and T. Bridges. 2002. Adoption of Conservation Production Systems in Two Ohio Watersheds: A Comparative Study. *Journal of Soil and Water Conservation* 57 (4): 229–235.

Nassauer, J.I. 1988. Landscape Care: Perceptions of Local People in Landscape Ecology and Sustainable Development. *Landscape and Land Use Planning*. Washington, DC: American Society of Landscape Architects.

———. 1989. Aesthetic Objectives for Agricultural Policy. *Journal of Soil and Water Conservation* 44: 384–387.

———. 1997a. Agricultural Landscapes in Harmony with Nature. In *Visions of American Agriculture*, edited by W. Lockeretz. Ames, IA: Iowa State University Press.

———. 1997b. The Landscape of American Agriculture: A Popular Image and a New Vision. In *Visions of American Agriculture*, edited by W. Lockeretz. Ames, IA: Iowa State University Press.

Nassauer, J.I., R.C. Corry, and R.M. Cruse. 2002. Alternative Future Landscape Scenarios: A Means to Consider Agricultural Policy. *Journal of Soil and Water Conservation* 57: 44A–53A.

Nassauer, J.I., and R.C. Corry. 2004. Using Normative Scenarios in Landscape Ecology. *Landscape Ecology* 19: 343–356.

Nassauer, J. I., and R.Westmacott. 1987. Progressiveness among Farmers as a Factor in Heterogeneity of Farmed Landscapes. In *Landscape Heterogeneity and Disturbance,* edited M.G. Turner. New York: Springer-Verlag.

Niemann, D. 1986. The Distribution of Orchids in Iowa. *Proceedings of the Iowa Academy of Science* 93(1): 24–34.

Niemann, D.A., and R. Q. Landers Jr. 1974. Forest Communities in Woodman Hollow State Preserve. *Proceedings of the Iowa Academy of Science* 81(4): 176–184.

Neumann, N.F., D. Smith, and M. Belosevic. 2005. Waterborne Disease: An Old Foe Re-emerging? *Journal of Environmental Engineering and Science* 4(3): 155–171.

Northwest Area Foundation. 1995. *A Better Row to Hoe—The Economic, Environmental, and Social Impact of Sustainable Agriculture.* Report Number: NAF17. Saint Paul, MN: Northwest Area Foundation.

Noss, R.F., and R.L. Peters. 1995. *Endangered Ecosystems: A Status Report of America's Vanishing Habitat and Wildlife.* Washington, DC: Defenders of Wildlife.

OECD (Organisation for Economic Co-operation and Development). Joint Working Party on Agriculture and Environment. 2002. *Agricultural Practices that Reduce Greenhouse Gas Emissions: Overview and Results of Survey Instrument.* June 2002. http://www.oecd.org/dataoecd/43/27/34421245.pdf (accessed December 6, 2005).

———. 2003. *Agricultural Policies in OECD Countries: A Positive Reform Agenda.* Paris: OECD.

Oglethorpe, D., N. Hanley, S. Hussain, and R. Sanderson. 2000. Modelling the Transfer of the Socio-economic Benefits of Environmental Management. *Environmental Modelling & Software* 15: 343–356.

Ouellet, M., J. Bonin, J. Rodrigue, J. DesGranges, and S. Lair. 1997. Hindlimb Deformities in Free-living Anurans from Agricultural Habitats. *Journal of Wildlife Diseases* 33:95–104.

Paolisso, M., and R.S. Maloney. 2000. Recognizing Farmer Environmentalism: Nutrient Runoff and Toxic Dinoflagellate Blooms in the Chesapeake Bay Region. *Human Organization* 59: 209–221.

Pearson, J.A., and M.J. Loeschke. 1992. Floristic Composition and Conservation Status of Fens in Iowa. *Journal of the Iowa Academy of Science* 99 (2–3): 41–52.

Peierls, B., N. Caraco, M. Pace, and J. Cole. 1991. Human Influence on River Nitrogen. *Nature* 350: 386–387.

Peles, J.D., D.R. Bowne, and G.W. Barrett. 1999. Influence of Landscape Structure on Movement Patterns of Small Mammals. In *Landscape Ecology of Small Mammals,* edited by G.W. Barret and J.D. Peles. New York: Springer.

Perkins, B.D., K. Lohman, E. Van Nieuwenhuyse, and J.R. Jones Jr. 1998. An Examination of Land Cover and Stream Water Quality among Physiographic Provinces of Missouri, U.S.A. *Verhandlungen der Internationalen Vereinigung für Theoretische und Angewandte Limnologie* 26:940–947.

Pew Center on Global Climate Change. 2005. *In Brief: Agriculture's Role in Addressing Climate Change.* http://www.pewclimate.org/docUploads/policy%5Finbrief%5Fag%2Epdf (accessed December 19, 2006).

Pimental, D., et al. 1997. Resources: Agriculture, the Environment, and Society. *Bioscience* 47 (2): 97–106.

Pioneer Hi-Bred International Incorporated. 2005. Related Product Information—Corn. *Insect Resistance Management: Preserving an Important Technology.* http://www.pioneer.com/products/US/sidebars/corn/insect_resistance.htm (accessed October 18, 2005).

PLANTS Database. 2007. USDA Natural Resources Conservation Service (NRCS). http://plants.usda.gov (accessed 1997).

Pope, S.E., L. Fahrig, and H.G. Merriam. 2000. Landscape Complementation and Metapopulation Effects on Leopard Frog Populations. *Ecology* 81:2498–2508.

Pratt, J.R., and J. Cairns Jr. 1992. Ecological Risks Associated with the Extinction of Species. In *Predicting Ecosystem Risk,* edited by J. Cairns Jr., B.R. Neiderlehner, and D.R. Orvos. Princeton,

NJ: Princeton University Scientific Publishing Co., Inc.

Pressey, R.L., C.J. Humphries, C.R. Margules, R.I. Van-Wright, and P.H. Williams. 1993. Beyond Opportunism: Key Principles for Systematic Reserve Selection. *Trends in Ecology and Evolution* 8: 124–128.

Puckett, L.J. 1994. *Nonpoint and Point Sources of Nitrogen in Major Watersheds of the United States.* Water Resources Investigations Report 94-4001.Washington, DC: USGS.

Rabalais, N.N. 2002. Nitrogen in Aquatic Systems. *Ambio* 31:102–112.

Rabalais, N.N., R. E. Turner, and D. Scavia. 2002. Beyond Science into Policy: Gulf of Mexico Hypoxia and the Mississippi River. *Bioscience* 52: 129–142.

Rabb, G.B., and T.B. Sullivan. 1995. Coordinating Conservation: Global Networking for Species Survival. *Biodiversity and Conservation* 4: 536–543.

Randall, G.W., and J.A. Vetsch. 2005. Nitrate Losses in Subsurface Drainage from a Corn-Soybean Rotation as Affected by Fall and Spring Application of Nitrogen and Nitrapyrin. *Journal of Environmental Quality* 34(2): 590–597.

Randall, G.W., D.R. Huggins, M.P. Russelle, D.J. Fuchs, W.W. Nelson, and J.L. Anderson. 1997. Nitrate Losses through Subsurface Tile Drainage in CRP, Alfalfa, and Row Crop Systems. *Journal of Environmental Quality* 26: 1240–1247.

Reeder, K.F., D.M. Debinski, and B.J. Danielson. 2005. Factors Affecting Butterfly Use of Filter Strips in Midwestern USA. *Agriculture, Ecosystems & Environment* 109: 40–47.

Reichelderfer, K., and W.G. Boggess. 1988. Government Decision Making and Program Performance: The Case of the Conservation Reserve Program. *American Journal of Agricultural Economics* 70: 1–11.

Reilly, J., et al. 2001. Agriculture and the Environment: Interactions with Climate from Agriculture: The Potential Consequences of Climate Variability and Change for the United States. In *US National Assessment of the Potential Consequences of Climate Variability and Change, US Global Change Research Program.* Cambridge University Press, New York, Chapter 5.

Relyea, R.A. 2005. The Impact of Insecticides and Herbicides on the Biodiversity and Productivity of Aquatic Communities. *Ecological Applications* 15 (2): 618–627.

Ribaudo, M., R. Horan, and M.E. Smith. 1999. Economics of Water Quality Protection from Nonpoint Sources: Theory and Practice. Agricultural Economics Report 728. Washington, DC: USDA.

Ribaudo, M., R. Heimlich, R. Claassen, and M. Peters. 2001. Least-cost Management of Nonpoint Source Pollution: Source Reduction vs. Interception Strategies for Controlling Nitrogen Loss in the Mississippi Basin. *Ecological Economics* 37: 183–197.

Richardson, M.S., and R.C. Gatti. 1999. Prioritizing Wetland Restoration Activity within a Wisconsin Watershed Using GIS Modeling. *Journal of Soil and Water Conservation* 54: 537–542.

Ries, L., and D.M. Debinski. 2001. Butterfly Responses to Habitat Edges in the Highly Fragmented Prairies of Central Iowa. *Journal of Animal Ecology* 70: 840–852.

Ries, L., D.M. Debinski, and M.L. Wieland. 2001. The Conservation Value of Roadside Prairie Restoration to Butterfly Populations. *Conservation Biology* 15(2): 401–411.

Risser, P.G. 1988. Diversity in and among Grasslands. In *Biodiversity,* edited by E.O. Wilson. Washington, DC: National Academy of Sciences.

Roberts, M. J., and N. Key. 2003. Who Benefits from Government Farm Payments? *Choices* 3rd qtr.: 7–14.

Robinson, C.A., R.M. Cruse, and M. Ghaffarzadeh. 1996. Cropping System and Nitrogen Effects on Mollisol Organic Carbon. *Soil Science Society of America Journal* 60(1): 264–269.

Roosa, D.M. 1981. Iowa Natural Heritage Preservation: History, Present Status and Future Challenges. *Proceedings of the Iowa Academy of Science* 88: 43–47.

Roosa, D.M., M.L. Leoschke, and L.J. Eilers. 1989. Distribution of Iowa's Endangered and Threatened Vascular Plants. *Iowa Department of Natural Resources.*

Rosenzweig, M.L. 1995. *Species Diversity in Space and Time.* Cambridge, England: Cambridge University Press.

———. 1999. Heeding the Warning in Biodiversity's Basic Law. *Science* 284: 276–277.

Runge, C.F. 1996. Agriculture and Environmental Policy: New Business or Business as Usual.

Working paper No. 1. The McKnight Foundation for Environmental Reform: The Next Generation Project, Yale Center for Environmental Law and Policy.

Russell, N.H. 1956. A Checklist of the Vascular Flora of Poweshiek County, Iowa. *Proceedings of the Iowa Academy of Science* 63: 161–176.

Rustigian, H., M. Santelmann, and N. Schumaker. 2003. Assessing the Potential Impacts of Alternative Landscape Designs on Amphibian Population Dynamics. *Landscape Ecology* 18: 65–81.

Ryan, R.L. 1998. Local Perceptions and Values for a Midwestern River Corridor. *Landscape and Urban Planning* 42: 225–237.

Ryan, R., D. Erickson, and R. DeYoung. 2003. Farmers' Motivations for Adopting Conservation Practices along Riparian Zones in a Mid-western Agricultural Watershed. *Journal of Environmental Planning and Management* 46 (1): 19–37.

Sala, O.E., et al. 2000. Global Biodiversity Scenarios for the Year 2100. *Science* 287: 1770–1774.

Salamon, S., R.L. Farnsworth, D.G. Bullock, and R. Yusuf. 1997. Family Factors Affecting Adoption of Sustainable Farming Systems. *Journal of Soil and Water Conservation* 52: 265–271.

Sandoz Agro Inc. 1993. The 1993 Sandoz National Agricultural Poll. Des Plaines, IL: Sandoz Agro, Inc.

Santelmann, M., et al. 2001. Applying Ecological Principles to Land-Use Decision Making in Agricultural Watersheds. In *Applying Ecological Principles to Land Management*, edited by V.H. Dale and R.A. Haeuber. New York: Springer-Verlag.

Santelmann, M., et al. 2004. Assessing Alternative Futures for Agriculture in Iowa, U.S.A. *Landscape Ecology* 19 (4): 357–374.

Santelmann, M., K. Freemark, J. Sifneos, and D. White. 2005. Assessing Effects of Alternative Agricultural Scenarios on Wildlife Habitat in Iowa, U.S.A. *Agriculture, Ecosystems & Environment* 113: 243–253.

Scavia, D., and S.B. Bricker. 2006. Coastal Eutrophication Assessment in the United States. *Biogeochemistry* 79: 187–208.

Scavia, D., D. Justic, and V. J. Bierman Jr. 2004. Reducing Hypoxia in the Gulf of Mexico: Advice from Three Models. *Estuaries* 27: 419–426.

Scavia, D., N. N. Rabalais, R. E. Turner, D. Justic, and W. J. Wiseman. 2003. Predicting the Response of Gulf of Mexico Hypoxia to Variations in Mississippi River Nitrogen Load. *Limnology and Oceanography* 48 (3): 951–956.

Schaumann, S. 1988. Scenic Value of Countryside Landscapes to Local Residents: A Whatcom County, Washington Case Study. *Landscape Journal* 7: 40–46.

Schepers, J.S., M.G. Moravek, E.E. Alberts, and K.D. Franks. 1991. Maize Production Impacts on Groundwater Quality. *Journal of Environmental Quality* 20: 12–16.

Schertz, L., and O. Doering. 1999. *The Making of the 1996 Farm Act*. Ames, IA: Iowa State University Press.

Schilling, K.E., and C.A. Thompson. 2000. Walnut Creek Watershed Monitoring Project, Iowa: Monitoring Water Quality in Response to Prairie Restoration. *Journal of the American Water Resources Association* 36 (5): 1101–1117.

Schlicht, D. W., and T. T. Orwig. 1998. The Status of Iowa's Lepidoptera. *Journal of the Iowa Academy of Science* 105: 82–88.

Schoon, B., and R. te Grotenhuis. 2000. Values of Farmers, Sustainability and Agricultural Policy. *Journal of Agricultural and Environmental Ethics* 12: 17–27.

Schnepf, R. 2006. *Agriculture-Based Renewable Energy Production*. Washington, DC: Library of Congress CRS.

Schrader, C.C. 1995. Rural Greenway Planning: The Role of Streamland Perception in Landowner Acceptance of Land Management Strategies. *Landscape and Urban Planning* 33: 375–390.

Schumaker, N.H. 1998. *A User's Guide to the PATCH Model*. EPA/600/R-98/135. U.S. Environmental Protection Agency, Environmental Research Laboratory, Corvallis, OR.

Schwartz, C.W., and E.R. Schwartz. 1981. *The Wild Mammals of Missouri*. Columbia, MO: University of Missouri Press.

Schwartz, M.W., and P.J. van Mantgem. 1997. The Value of Small Preserves in Chronically Fragmented Landscapes. In *Conservation in Highly Fragmented Landscapes*, edited by M.W. Schwartz. New York: Chapman & Hall.

Scott, J.A. 1986. *The Butterflies of North America*. Stanford, CA: Stanford University Press.

Seale, D.B. 1982. Physical Factors Influencing Oviposition by the Wood Frog, *Rana sylvatica*, in Pennsylvania. *Copeia* 1982: 627–635.

Selman, P., and N. Doar. 1992. An Investigation of the Potential for Landscape Ecology to Act as a Basis for Rural Land Use Plans. *Journal of Environmental Management* 35: 281–299.

Semlitsch, R.D. 1983. Structure and Dynamics of Two Breeding Populations of the Eastern Tiger Salamander, *Ambystoma tigrinum*. *Copeia* 1983: 608–616.

———. 1987. Relationship of Pond Drying to the Reproductive Success of the Salamander *Ambystoma talpoideum*. *Copeia* 1987:61–69.

Sharpley, A.N., and J.R. Williams. 1990. *EPIC—Erosion Productivity Impact Calculator: Vol. I. Model Documentation*. Technical Bulletin No. 1768. Washington, DC: USDA.

Shepard, R. 2005. Nutrient Management Planning: Is It the Answer to Better Management? *Journal of Soil and Water Conservation* 60(4): 171–76.

Shepherd, S., and D.M. Debinski. 2005a. Evaluation of Isolated and Integrated Prairie Reconstructions as Habitat for Prairie Butterflies. *Biological Conservation* 126: 51–61.

———. 2005b. Reintroduction of Regal Fritillary (*Speyeria idalia*) to a Restored Prairie. *Ecological Restoration* 23(4): 243–249.

Shoup, M.J. 1999. Agricultural Runoff of Nutrients and Indicator Organisms in Buck Creek, Iowa. Master of Science Thesis, Department of Civil and Environmental Engineering, University of Iowa, Iowa City, IA.

Sih, A., P. Crowley, M. McPeek, J. Petranka, and K. Strohmeier. 1985. Predation, Competition, and Prey Communities: A Review of Field Experiments. *Annual Review of Ecology and Systematics* 16: 269–311.

Silva, M., and J.A. Downing. 1995. *CRC Handbook of Mammalian Body Masses*. Boca Raton, FL: CRC Press.

Skelly, D.K. 1995. A Behavioral Trade-off and Its Consequences for the Distribution of *Pseudacris* Treefrog Larvae. *Ecology* 76:150–164.

———. 1996. Pond Drying, Predators, and the Distribution of *Pseudacris* Tadpoles. *Copeia* 1996: 599–605.

Skibbe, J.D.D. 2005. Butterfly Community Composition in Fragmented Habitats: Effects of Patch Shape and Spatial Scale. M.S. Thesis, Iowa State University.

Smart, M.M., J.R. Jones, and J.L. Sebaugh. 1985. Stream-Watershed Relations in the Missouri Ozark Plateau Province. *Journal of Environmental Quality* 14: 77–82.

Smith, M. 2000. *Agricultural Resources and Environmental Indicators: Land Retirement*. Washington, DC: USDA ERS.

Smith, O.H., G.W. Petersen, and B.A. Needelman. 2000. Environmental Indicators of Agroecosystems. *Advances in Agronomy* 69: 75–97.

Smith, R.B.W. 1995. The Conservation Reserve Program as a Least-Cost Land Retirement Mechanism. *American Journal of Agricultural Economics* 77 (1): 93–105.

Snyder C.S., and T.W. Bruulsema. 2002. *Nutrients and Environmental Quality, Plant Nutrient Use in North American Agriculture*. PPI/PPIC/FAR Technical Bulletin 2002-1, May 2002, Potash & Phosphate Institute, Norcross, GA. http://www.ppi-ppic.org (accessed December 19, 2006).

Soulé, M.E. 1991. Conservation Tactics for a Constant Crisis. *Science* 253: 744–750.

Starfield, A. 1997. A Pragmatic Approach to Modeling for Wildlife Management. *Journal of Wildlife Management* 61: 261–270.

Stohlgren, T.J., G.W. Chong, M.A. Kalkhan, and L.D. Schell. 1997. Rapid Assessment of Plant Diversity Patterns: A Methodology for Landscapes. *Ecological Monitoring and Assessment* 48: 25–43.

Stone, L., and A. Roberts. 1992. Conditions for a Species to Gain Advantage from the Presence of Competitors. *Ecology* 72: 1964–1972.

Stout, W. L., S. L. Fales, L.D. Muller, R.R. Schnabel, G.F. Elwinger, and S.R. Weaver. 2000. Assessing the Effect of Management Intensive Grazing on Water Quality in the Northeast US. *Journal of Soil and Water Conservation* 55: 238–243.

Sutcliffe, O. L., V. Bakkestuen, G. Fry, and O.E. Stabbetorp. 2003. Modeling the Benefits of Farmland Restoration: Methodology and Application to Butterfly Movement. *Landscape and Urban Planning* 63: 15–31.

Suter, W. 1998. Involving Conservation Biology in Biodiversity Strategy and Action Planning. *Biological Conservation* 83: 235–237.

Sullivan, P., et al. 2004. *The Conservation Reserve Program: Economic Implications for Rural America.* Washington, DC: USDA ERS.

Swaffield, S.R. 1998. Contextuality in Policy Discourse: A Case Study of Language Use Concerning Resource Policy in the New Zealand High Country. *Policy Sciences* 31:199–224.

Taylor, P., L. Fahrig, K. Henein, and H.G. Merriam. 1993. Connectivity Is a Vital Element of Landscape Structure. *Oikos* 68: 571–573.

Tewksbury, J.J., et al. 2002. Corridors Affect Plants, Animals, and Their Interactions in Fragmented Landscapes. *Proceedings of the National Academy of Sciences of the United States of America* 99: 12923–12926.

Thomas, G.W., K.L. Wells, and L. Murdock. 1981. Fertilization and Liming. In *No Tillage Research: Research Reports and Reviews*, edited by R.E. Phillips, G.W. Thomas, and R.L. Blevins. Lexington, KY: University of Kentucky College of Agriculture and Agricultural Experiment Station.

Thompson, J. 1992. *Prairies, Forests and Wetlands: The Restoration of Natural Landscape Communities in Iowa.* Iowa City, IA: University of Iowa Press.

Thompson, L.M. 1957. *Soils and Soil Fertility.* New York: McGraw-Hill Book Co.

Tilman, D., et al. 2001. Forecasting Agriculturally Driven Global Environmental Change. *Science* 292: 281–284.

TNC (The Nature Conservancy) and NatureServe. 2003a. Fact Backgrounder: Quick Facts from the Upper Mississippi River Basin Report. http://nature.org/wherewework/northamerica/states/missouri/files/umr.pdf (accessed December 19, 2006).

———. 2003b. Report: Conservation Priorities for Freshwater Biodiversity in the Upper Mississippi River Basin. http://conserveonline.org/docs/2003/08/UMRB_report.pdf (accessed December 19, 2006).

Traore, N., R. Landry, and N. Amara. 1998. On-farm Adoption of Conservation Practices: The Role of Farm and Farmer Characteristics, Perceptions, and Health Hazards. *Land Economics* 74: 114–127.

Turner, R.E. 2005. Nitrogen and Phosphorus Concentration and Retention in Water Flowing over Freshwater Wetlands. In *Ecology and Management of Bottomland Hardwood Systems: The State of Our Understanding*, edited by L. Fredrickson, S.L. King, and R.M. Kaminski. Columbia, MO: University of Missouri Press.

Turner, R.E., and N.N. Rabalais. 1991. Changes in the Mississippi River This Century: Implications for Coastal Food Webs. *BioScience* 41: 140–147.

———. 1994. Coastal Eutrophication Near the Mississippi River Delta. *Nature* 368: 619–621.

———. 2003. Linking Landscape and Water Quality in the Mississippi River Basin for 200 Years. *BioScience* 53: 563–572.

———. 2004. Suspended Sediment, C, N, P, and Si Yields from the Mississippi River Basin. Hydrobiologia 511: 79–89.

Turner, R.E., N.N. Rabalais, and D. Justic. 2006. Predicting Summer Hypoxia in the Northern Gulf of Mexico: Riverine N, P and Si Loading. *Marine Pollution Bulletin* 52(2): 139–148.

Turner, R.E., N.N. Rabalais, E.M. Swenson, M. Kasprzak, and T. Romaire. 2005. Summer Hypoxia in the Northern Gulf of Mexico and Its Prediction from 1978 to 1995. *Marine Environmental Research* 59(1): 65–77.

Turner, R.E., N. Qureshi, N.N. Rabalais, Q. Dortch, D. Justi, R. Shaw, and J. Cope. 1998. Fluctuating Silicate: Nitrate Ratios and Coastal Plankton Food Webs. *Proceedings of the National Academy of Sciences of the United States of America* 95: 13048–13051.

Turner, R.E., D. Stanley, D. Brock, J. Pennock, and N.N. Rabalais. 2000. A Comparison of Independent N-Loading Estimates for U.S. Estuaries. In *Nitrogen Loading in Coastal Water Bodies. An Atmospheric Perspective (Coastal and Estuarine Sciences)*, edited by R.A. Valigura, R.B. Alexander, M.S. Castro, T.P. Meyers, H.W. Paerl, P.E. Stacey, and R.E. Turner. Washington, DC: American Geophysical Union.

Tyler, D.D., and G.W. Thomas. 1979. Lysimeter Measurement of Nitrate and Chloride Losses from Conventional and No-tillage Corn. *Journal of Environmental Quality* 6: 63–66.

Ubelaker, D.H. 1992. The Sources and Methodology for Mooney's Estimates of North American Indian Populations. In *The Native Population of the Americas in 1492*, edited by W.H. Denevan. Madison, WI: Wisconsin Press.

UNEP (United Nations Environmental Programme). 2003. *Global Environmental Outlook 2003*. http://www.unep.org/geo/ (accessed September 14, 2006).

Urban, M. 2005. Values and Ethical Beliefs Regarding Agricultural Drainage in Central Illinois, USA. *Society and Natural Resources* 18: 173–189.

USDA. 1981. *A Time to Choose: Summary Report on the Structure of Agriculture*. Washington, DC: USDA.

———. 2006. Conservation and the Environment: USDA 2007 Farm Bill Theme Paper. USDA.

USDA ERS. 2006. Farm and Commodity Policy: Government Payments and the Farm Sector, Briefing Room Report. USDA Economic Research Service. February 24, 2006. http://www.ers.usda.gov/Briefing/FarmPolicy/gov-pay.htm (accessed December 19, 2006).

USDA ERS. 2006. Farm Resources Regions Agricultural Information Bulletin Number 760. http://www.ers.usda.gov/publications/aib760/aib-760.pdf (accessed December 19, 2006).

USDA FSA (Farm Services Agency). 2002. Conservation Reserve Program Reports. http://www.fsa.usda.gov/dafp/cepd/16CRP/st1-4.pdf (accessed January 16, 2002, no longer at this address). Updated URL: http://www.fsa.usda.gov/FSA/webapp?area=home&subject=copr&topic=crp-rt (accessed December 19, 2006).

USDA NASS (National Agricultural Statistics Service). 1999. *1997 Census of Agriculture: Iowa State and County Data*. Report Number: AC97-A-15. Washington, DC: USDA.

———. Census of Agriculture. 2002. http://www.nass.usda.gov/Census_of_Agriculture/index.asp (accessed December 19, 2006).

———. 2004. *2002 Census of Agriculture: Iowa State and County Data*. Report Number: AC97-A-15. Washington, DC: USDA.

———. 2007. Prospective Plantings Report. Washington, DC: USDA.

USDA NRCS (Natural Resources Conservation Service). 1997. *Profitable Pastures*. Des Moines, IA: USDA NRCS.

———. 2006. Conservation and the Environment: USDA 2007 Farm Bill Theme Paper. Washington, DC: USDA. http://www.usda.gov/documents/FarmBill07consenv.pdf (accessed December 10, 2006).

———. 2007. The PLANTS database. Baton Rouge, LA: National Plant Data Center. http://plants.usda.gov (accessed March 12, 2007).

USDA SCS (Soil Conservation Service). 1994. Conservation Choices. St. Paul, MN: U.S. Government Printing Office, 546–791.

U.S. EPA (Environmental Protection Agency). 1998. *National Water Quality Inventory: 1996 Report to Congress*. EPA841-F-97-003. Washington, DC: U.S. EPA Office of Water.

———. 2001. *Action Plan for Reducing, Mitigating, and Controlling Hypoxia in the Northern Gulf of Mexico*. Task force paper, Mississippi River/Gulf of Mexico Watershed Nutrient Task Force. Washington, DC: U.S. EPA Office of Wetlands, Oceans, and Watersheds. http://www.epa.gov/msbasin/taskforce/pdf/actionplan.pdf (accessed December 17, 2006).

———. 2002. 2000 National Water Quality Inventory. EPA841-R-02-001.Washington, DC: U.S. EPA Office of Water. http://www.epa.gov/305b/2000report (accessed December 19, 2006).

———. 2005. National Center for Environmental Research, Research Opportunities Archives. http://es.epa.gov/ncer/rfa/archive/grants/ (accessed November 16, 2005).

———. 2006. Biodiversity Data. Corvallis, OR: U.S. EPA Western Ecology Division. http://www.epa.gov/wed/pages/staff/white/getbiod.htm (accessed December 19, 2006).

Vaché, K.B., J.M. Eilers, and M.V. Santelmann. 2002. Water Quality Modeling of Alternative Agricultural Scenarios for the U.S. Corn Belt. *Journal of the American Water Resources Association* 38(3): 773–787.

VanBreemen N., et al. 2002. Where Did All the Nitrogen Go? Fate of Nitrogen Inputs to Large Watersheds in the Northeastern USA. *Biogeochemistry* 57:267–293.

van den Berg, A. E., C.A.J. Vlek, and J.F. Coeterier. 1998. Group Differences in the Aesthetic Evaluation of Nature Development Plans: A Multilevel Approach. *Journal of Environmental Psychology* 18(2): 141–157.

Van der Linden, P., and D.R. Farrar. 1993. *Forest and Shade Trees of Iowa.* Ames, IA: Iowa State University Press.

Vitousek, P.M., H.A. Mooney, J. Lubchenco, and J.M. Melillo. 1997a. Human Domination of Earth's Ecosystems. *Science* 277: 494–499.

Vitousek, P.M., et al. 1997b. Human Alteration of the Global Nitrogen Cycle: Sources and Consequences. *Ecological Applications* 7: 737–750.

Vogel, S. 1996. Farmers' Environmental Attitudes and Behavior: A Case Study for Austria. *Environment and Behavior* 28: 591–613.

Vos, C.C., and A.H.P. Stumpel. 1995. Comparison of Habitat-Isolation Parameters in Relation to Fragmented Distribution Patterns in the Tree Frog. *Landscape Ecology* 11:203–214.

Vyn, Tony. 2007. Meeting the Ethanol Demand: Consequences and Compromises Associated with More Corn on Corn in Indiana. Lafayette, IN: Purdue Cooperative Extension Service.

Waide, J.B., and J.L. Hatfield. 1995. *Preliminary MASTER Assessment of the Impacts of Alternative Agricultural Management Practices on Ecological and Water Resource Attributes of Walnut Creek Watershed, Iowa.* Report Number: 07IMG1718: 68-CO-0050. Little Rock, AR: FTN Associates.

Wake, D.B. 1991. Declining Amphibian Populations. *Science* 253:860.

Walter, G. 1997. Images of Success: How Illinois Farmers Define the Successful Farmer. *Rural Sociology* 62: 48–68.

Weber, R. P. 1990. *Basic Content Analysis.* Beverly Hills, CA: Sage Publications.

Weibull, A.C., and O. Ostman. 2003. Species Composition in Aggregate Ecosystems: The Effect of Landscape, Habitat, and Farm Management. *Basic and Applied Ecology* 4: 349–361.

Westerman, P.R., A. Hofman, L.E.M. Vet, and W. van der Werf. 2003a. Relative Importance of Vertebrates and Invertebrates in Epigeaic Weed Seed Predation in Organic Cereal Fields. *Agriculture, Ecosystems and the Environment* 95: 417–425.

Westerman, P.R., J.S. Wes, M.J. Kropff, and W. van der Werf. 2003b. Annual Losses of Weed Seeds Due to Predation in Organic Cereal Fields. *Journal of Applied Ecology* 40: 824–836.

Westmacott, R., and T. Worthington. 1984. *Agricultural Landscapes: A Second Look.* Report Number: CCP 168. Countryside Commission.

Weyer, P. 2001. Nitrate in Drinking Water and Human Health. Center for the Health Effects of Environmental Contamination. University of Iowa. Paper prepared for the University of Illinois Urbana-Champaign Agriculture Safety and Health Conference held in March 2001.

White, D., E. M. Preston, K. E. Freemark, and A. R. Kiester. 1999. A Hierarchical Framework for Conserving Biodiversity. In *Landscape Ecological Analysis: Issues and Applications*, edited by J.M. Klopatek and R.H. Gardner. New York: Springer.

White, D., et al. 1997. Assessing Risks to Wildlife Habitat from Future Landscape Change. *Conservation Biology* 11: 349–360.

Whittaker, R.H. 1970. *Communities and Ecosystems.* London: Macmillan.

Wilbur, H.M. 1987. Regulation of Structure in Complex Systems: Experimental Temporary Pond Communities. *Ecology* 68:1437–1452.

Williams, J.R., C.A. Jones, and P.T. Dyke. 1988. *EPIC, the Erosion Productivity Index Calculator, Model Documentation* Vol. 1. Temple, TX: USDA ARS.

Willits, F.K., and A.E. Luloff. 1995. Urban Residents' Views of Rurality and Contacts with Rural Places. *Rural Sociology* 60: 454–466.

Willock, J., et al. 1999. The Role of Attitudes and Objectives in Farmer Decision Making: Business and Environmentally-oriented Behaviour in Scotland. *Journal of Agricultural Economics* 50: 286–303.

Wilson, G.A. 1997. Factors Influencing Farmer Participation in the Environmentally Sensitive Areas Scheme. *Journal of Environmental Management* 50: 67–93.

Withers, M., M.W. Palmer, G.L. Wade, P.S. White, and P.R. Neal. 2003. Changing Patterns in the Number of Species in North American Floras. Chapter 4 in *The Land Use History of North America*. USGS. Last updated: Thursday, November 20, 2003 15:21:37 MST. http://biology.usgs.gov/luhna/chap4.html (accessed October 24, 2005).

Wolf, A.T. 1995. Rural Nonpoint Source Pollution Control in Wisconsin: The Limits of a Voluntary Program? *Water Resources Bulletin* 31: 1009–1022.

Woltemade, C. J. 2000. Ability of Restored Wetlands to Reduce Nitrogen and Phosphorus Concentrations in Agricultural Drainage Water. *Journal of Soil and Water Conservation* 55: 303–309.

Womach, J., et al. 2006. *Previewing the Farm Bill*. Report number RL33037. Washington, DC: Library of Congress CRS. January 30.

Woodward, R., and Y.S. Wui. 2001. The Economic Value of Wetland Services: A Meta-analysis. *Ecological Economics* 37: 257–270.

Wragg, A. 2000. Towards Sustainable Landscape Planning: Experiences from the Wye Valley Area of Outstanding Natural Beauty. *Landscape Research* 25(2): 183–200.

Wu, J., and B.A. Babcock. 1999. Metamodeling Potential Nitrate Water Pollution in the Central United States. *Journal of Environmental Quality* 28: 1916–1928.

Wu, J., R. Adams, C.L. Kling, and K. Tanaka. 2004. From Microlevel Decisions to Landscape Changes: An Assessment of Agricultural Conservation Policies. *American Journal of Agricultural Economics* 86(1): 26–41.

Zedler, J. 2003. Wetlands at Your Service: Reducing Impacts of Agriculture at the Watershed Scale. *Frontiers in Ecology and the Environment* 1(2): 65–72.

Zinn, J., and T. Cowan. 2006. *Agriculture Conservation Programs: A Scorecard*. CRS Report RS22243, March 17, 2006. Updated CRS Report RL32940, April 10, 2006. Washington, DC: Library of Congress CRS.

Zucker, L.A., and L.C. Brown (eds.). 1998. *Agricultural Drainage: Water Quality Impacts and Subsurface Drainage Studies in the Midwest*. OSU Extension Bulletin number 871. Columbus, OH: Ohio State University.

# Index